Springer Proceedings in Energy

The series Springer Proceedings in Energy covers a broad range of multidisciplinary subjects in those research fields closely related to present and future forms of energy as a resource for human societies. Typically based on material presented at conferences, workshops and similar scientific meetings, volumes published in this series will constitute comprehensive state-of-the-art references on energy-related science and technology studies. The subjects of these conferences will fall typically within these broad categories:

- Energy Efficiency
- Fossil Fuels
- Nuclear Energy
- Policy, Economics, Management & Transport
- Renewable and Green Energy
- Systems, Storage and Harvesting
- Materials for Energy

eBook Volumes in the Springer Proceedings in Energy will be available online in the world's most extensive eBook collection, as part of the Springer Energy eBook Collection. To submit a proposal or for further inquiries, please contact the Springer Editor in your region:

Kamiya Khatter (India)
Email: kamiya.khatter@springer.com
Loyola D'Silva (All other countries)
Email: loyola.dsilva@springer.com

Mahmod Bin Othman ·
Samsul Ariffin Abdul Karim ·
Cecilia Devi Wilfred · Kean Chuan Lee ·
Rajalingam Sokkalingam
Editors

Proceedings of the 1st International Conference of New Energy

ICNE 2022, 1-2 Dec, Sarawak, Malaysia

Springer

Editors
Mahmod Bin Othman
Universiti Teknologi PETRONAS
Perak, Malaysia

Cecilia Devi Wilfred
Universiti Teknologi PETRONAS
Perak, Malaysia

Rajalingam Sokkalingam
Universiti Teknologi PETRONAS
Perak, Malaysia

Samsul Ariffin Abdul Karim
Software Engineering Programme, Faculty of Computing and Informatics
Universiti Malaysia Sabah
Kota Kinabalu, Malaysia

Data Technologies and Applications (DaTA) Research Lab, Faculty of Computing and Informatics
Universiti Malaysia Sabah
Kota Kinabalu, Malaysia

Kean Chuan Lee
Universiti Teknologi PETRONAS
Perak, Malaysia

ISSN 2352-2534 ISSN 2352-2542 (electronic)
Springer Proceedings in Energy
ISBN 978-981-99-0858-5 ISBN 978-981-99-0859-2 (eBook)
https://doi.org/10.1007/978-981-99-0859-2

This Springer imprint is published by the registered company Springer Nature Singapore Pte Ltd.
The registered company address is: 152 Beach Road, #21-01/04 Gateway East, Singapore 189721, Singapore

Foreword

The Department of Fundamental and Applied Sciences, Universiti Teknologi PETRONAS (UTP) is pleased to announce the 1st International Conference on New Energy (ICNE2022) under the banner of World Engineering, Science & Technology Congress (ESTCON 2022). The conference was held in Borneo Convention Centre, Kuching, Sarawak, Malaysia from 1st–2nd December 2022.

With the theme of "Shaping Hydrogen Economy towards a sustainable Future", ICNE conference aims to bring together academics, scientists, and industrialists from all around the world for knowledge sharing, exchange of ideas, and to present research findings in all aspects of hydrogen exploration. The conference has provided opportunities for the delegates to exchange new ideas and innovations, to establish research initiatives or business relations, and find global partners for future collaboration.

November 2022

Assoc. Prof. Dr. Maizatul Shima Binti Shaharun
Chair
Fundamental of Applied Sciences Department
Universiti Teknologi PETRONAS
Seri Iskandar, Perak Darul Ridzuan, Malaysia

Preface

The Fundamental and Applied Sciences Department, Universiti Teknologi PETRONAS is honoured to host the ICNE2022. Special appreciation and thank you to all the reviewers for providing their expertise in improving the quality of the papers, without which this book will not be possible. The second editor is gratefully to the Faculty of Computing and Informatics, Universiti Malaysia Sabah for the very supportive during the completion of the proceeding compilation. Finally, thank you to Springer for publishing the conference in Springer Proceedings in Energy.

Seri Iskandar, Malaysia	Mahmod Bin Othman
Kota Kinabalu, Malaysia	Samsul Ariffin Abdul Karim
Seri Iskandar, Malaysia	Cecilia Devi Wilfred
Seri Iskandar, Malaysia	Kean Chuan Lee
Seri Iskandar, Malaysia	Rajalingam Sokkalingam
November 2022	

Editorial for the Special Issue "Hydrogen as the New Sustainable Renewable Energy"

Fossil fuel consumption is associated with progressive release of greenhouse gases, which affects our climate. In recent years the global push towards a hydrogen economy has gained momentum as the focus on climate change has intensified. The global momentum for hydrogen is growing as we shift away from fossil fuels. Hydrogen is a lesser-known resource, despite being the most abundant clean fuel in the universe. Hydrogen is a clean fuel that, when consumed in a fuel cell, produces only water. Hydrogen can be produced from a variety of domestic resources, such as natural gas, nuclear power, biomass, and renewable power like solar and wind. These qualities make it an attractive fuel option for transportation and electricity generation applications. It can be used in cars, in houses, for portable power, and in many more applications.

The terms "grey", "blue", and "green" are being associated when describing hydrogen technologies. Green hydrogen is defined as hydrogen produced by splitting water into hydrogen and oxygen using renewable electricity. Grey hydrogen is traditionally produced from methane (CH_4), split with steam into CO_2 and H_2. Grey hydrogen has increasingly been produced also from coal, with significantly higher CO_2 emissions per unit of hydrogen produced, so much that is often called brown or black hydrogen instead of grey. Blue hydrogen follows the same process as grey, with the additional technologies necessary to capture the CO_2 produced when hydrogen is split from methane and stored for a long term.

This book will highlight the current trend studies on fundamental studies of hydrogen technologies to application of hydrogen as the new sustainable renewable energy. The coverage is the following areas: hydrogen production, its storage and transportation, and hydrogen utilisation. It is hoped that this work will contribute in making green hydrogen competitive and ready for a scale up in the 2030s, towards the objective of reaching net zero emissions by 2050.

Chapter Introduction

This book highlights the latest advancement in hydrogen exploration. With the theme of "Shaping Hydrogen Economy towards a sustainable Future", ICNE 2022 brings together leading experts, scientific communities, and industrialists working in the field of applied sciences and mathematics from all over the world to share the most recent developments and cutting-edge discoveries addressing all aspects of hydrogen exploration. The contents of this book are useful for researchers, students, and industrial practitioners in the areas of Mathematics, Physics, and Chemistry as most of the topics are in line with the sustainable future.

List of Reviewers

1. Dr. Zulkifli Merican Aljunid Merican
2. Associate Prof Dr. Noor Asmawati Zabidi
3. Associate Prof Dr. Anita Ramli
4. Dr. Hamzah Sakidin
5. Mr. Mohd Faisal Taha
6. Dr. Dennis Chuan Ching
7. Dr. Lim Jun Wei
8. Dr. Lee Kean Chuan
9. Associate Prof Dr. Ibrahima Faye
10. Associate Prof Dr. Ts. Dr. Mahmod Othman
11. Dr. Nooraini Zainuddin
12. Dr. Normawati Yunus
13. Associate Prof Dr. Hanita Daud
14. Dr. Khairul Azhar Zahri
15. Associate Prof Dr. Cecilia Wilfred
16. Dr. Mohana Sundaram Muthuvalu

17. Dr. Muhammad Fadhlullah Abd Shukur
18. Dr. Noraini Ghani
19. Dr. Khe Cheng Seong
20. Dr. Rajalingam Sokkalingam
21. Dr. Beh Hoe Guan

Contents

Editors and Contributors

About the Editors

Mahmod Bin Othman is currently an Associate Professor at the Universiti Teknologi PETRONAS, Seri Iskandar, Perak, MALAYSIA since March 2016. Before that, he was an Associate Professor at the Universiti Teknologi MARA, Perlis, MALAYSIA. He has a total of twenty-three years (23) of working experience in education industries as an academician where twenty-one (21) years with Universiti Teknologi MARA and another six (6) years with Universiti Teknologi PETRONAS until present. He is a Professional Technologist registered with Malaysia Board of Technologists (MBOT). He is a Member of International Association of Survey Statisticians, Malaysian Mathematical Sciences Society/Ahli Persatuan Sains dan Matematik (PERSAMA), Management Science/Operation Research Society Malaysia, and HRDF Certified Trainer. His research interests span the fields of Fuzzy Mathematics, Artificial Intelligent, Optimization, and Decision Making.

Samsul Ariffin Abdul Karim is an Associate Professor with Software Engineering Programme, Faculty of Computing and Informatics, Universiti Malaysia Sabah (UMS), Malaysia. He is Core Member of Data Technologies and Applications (DaTA) Research Lab, Faculty of Computing and Informatics, Universiti Malaysia Sabah. He obtained his Ph.D. in Mathematics from Universiti Sains Malaysia (USM) majoring in Computer Aided Geometric Design (CAGD). He is a Professional Technologist registered with Malaysia Board of Technologists (MBOT), No. Perakuan PT21030227. His research interest includes numerical analysis, fast algorithm for ODE and PDE, machine learning, approximation theory, optimisation, image processing, spline function, wavelets, and engineering education. He has published more than 170 papers in Journals and Conferences including three Edited Conferences Volumes and 80 book chapters. He was the recipient of Effective Education Delivery Award and Publication Award (Journal & Conference Paper), UTP Quality Day 2010, 2011, and 2012, respectively. He was Certified WOLFRAM Technology Associate, Mathematica Student Level. He also has published ten books with Springer Publishing including five books with Studies in Systems, Decision and Control (SSDC) series, two books with Taylor and Francis/CRC Press, one book with IntechOpen, and one book with UTP Press. Recently he has received Book Publication Award in UTP Quality Day 2020 for the book Water Quality Index (WQI) Prediction Using Multiple Linear Fuzzy Regression: Case Study in Perak River, Malaysia, which was published by SpringerBriefs in Water Science and Technology in 2020.

Dr. Cecilia Devi Wilfred is a faculty staff at Universiti Teknologi PETRONAS in Malaysia. She pursued her Ph.D. in Organic Chemistry with The University of York, in UK and did her postdoctoral at Queens University of Belfast. Her current research interests are mainly focused on organic synthesis, Ionic Liquids and their applications, and Green Chemistry.

Dr. Kean Chuan Lee is working as a Senior Lecturer in Universiti Teknologi PETRONAS under Fundamental and Applied Sciences Department teaching courses in Applied Physics programs which includes Renewable Energy, Renewable Energy Conversion, and Nanoelectronics. He pursued his Ph.D. at Tokyo Institute of Technology. He is a Professional Technologist registered with Malaysia Board of Technologists (MBOT). His current research interests are mainly focused on nanomaterials application in Oil and Gas industry.

Rajalingam Sokkalingam is a Senior Lecturer in the Department of Fundamental and Applied Sciences, Universiti Teknologi PETRONAS, Malaysia. He received an M.Sc. in Industrial Technology and Ph.D. degree in Mathematics. He has total 27 years of working experience with 12 years in manufacturing industries in engineering field and another 15 years in education industries as an academician. The subjects that he has taught range from the first year undergraduates (Mathematics and Statistics) up to the postgraduate level (Risk, Project, and Cost Management). He managed to present his research work at a number of national and international conferences, as well as published some results in national and international journals. He possesses good experience in computer laboratory sessions and is very familiar with some related mathematics and statistical software, such as Maple, SPSS, and Expert Design.

Contributors

Nor Ain Fathihah Binti Abdullah Department of Fundamental and Applied Sciences, Universiti Teknologi PETRONAS, Bandar Seri Iskandar, Perak, Malaysia

Sarah Najm Al-Challabi Mechanical Engineering Department, Universiti Teknologi PETRONAS, Tronoh, Malaysia

Thar Albarody Mechanical Engineering Department, Universiti Teknologi PETRONAS, Tronoh, Malaysia

Ana Hasrinatullina Bt M Basri PETRONAS Carigali Sendirian Berhad, Kuala Lumpur, Malaysia

Gerard Ang Electrical Engineering Department, School of Graduate Studies, Mapua University, Manila, Philippines

Fatima Musa Ardo HICoE-Centre for Biofuel and Biochemical Research (CBBR), Institute of Self-Sustainable Building (ISB), Department of Fundamental and Applied Sciences, Universiti Teknologi PETRONAS, Perak Darul Ridzuan, Malaysia

Mahmood Riyadh Atta Department of Fundamental and Applied Sciences, Universiti Teknologi PETRONAS, Seri Iskandar, Perak, Malaysia

Venugopal Balakrishnan Institute for Research in Molecular Medicine (INFORMM), Universiti Sains Malaysia, USM, George Town, Penang, Malaysia

Nor Hafizah Berahim Center of Contaminant Control and Utilization (CenCoU), Institute of Contaminant Management for Oil and Gas, Department of Fundamental and Applied Sciences, Universiti Teknologi PETRONAS, Seri Iskandar, Perak, Malaysia;
PETRONAS Research Sdn.Bhd., Kajang, Selangor, Malaysia

Usman Ghani State Key Laboratory of Metal Matrix Composites, Shanghai Jiao Tong University, Shanghai, PR China

Beh Hoe Guan Department of Fundamental and Applied Sciences, Universiti Teknologi PETRONAS, Seri Iskandar, Perak, Malaysia

Abdurrashid Haruna Department of Fundamental and Applied Sciences, Universiti Teknologi PETRONAS, Seri Iskandar, Perak Darul Ridzuan, Malaysia;
Department of Chemistry, Ahmadu Bello University, Zaria, Nigeria;
Institute of Contaminant Management, Universiti Teknologi PETRONAS, Seri Iskandar, Perak Darul Ridzuan, Malaysia

Almila Hassan Department of Fundamental and Applied Sciences, Universiti Teknologi PETRONAS, Seri Iskandar, Perak, Malaysia

Hitoshi Inokawa Division of Applied Chemistry, Gradated School of Engineering, Sojo University, Ikeda, Nishi-ku, Kumamoto, Japan

Junji Inukai University of Yamanashi, Kofu Yamanashi, Japan

Mohamad Adil Iman Bin Ishak Department of Fundamental and Applied Sciences, Universiti Teknologi PETRONAS, BandarSeri Iskandar, Perak, Malaysia

Khairulazhar Jumbri Department of Fundamental and Applied Sciences, Universiti Teknologi PETRONAS, Seri Iskandar, Perak, Malaysia

Khairulazhar Bin Jumbri Department of Fundamental and Applied Sciences, Universiti Teknologi PETRONAS, BandarSeri Iskandar, Perak, Malaysia

Katsuki Kusakabe Division of Applied Chemistry, Gradated School of Engineering, Sojo University, Ikeda, Nishi-ku, Kumamoto, Japan

Jun Wei Lim HICoE-Centre for Biofuel and Biochemical Research (CBBR), Institute of Self-Sustainable Building (ISB), Department of Fundamental and Applied Sciences, Universiti Teknologi PETRONAS, Perak Darul Ridzuan, Malaysia

M Faris B M Shah PETRONAS Carigali Sendirian Berhad, Kuala Lumpur, Malaysia

Ronnel Delos Santos Magbitang School of Graduate Studies, Mapua University, Manila, Philippines

Aishah Mahpudz Division of Applied Chemistry, Gradated School of Engineering, Sojo University, Ikeda, Nishi-ku, Kumamoto, Japan

M. D. Maksudur Rahman Khan Faculty of Chemical and Natural Resources Engineering, Universiti Malaysia Pahang, Gambang, Pahang, Malaysia;
Petroleum and Chemical Engineering, Faculty of Engineering, Universiti Teknologi Brunei, Bandar Seri Begawan, Brunei Darussalam

Pravin Mariappan Mechanical Engineering Department, Universiti Teknologi PETRONAS, Tronoh, Malaysia

Zulkifli Merican Aljunid Merican Department of Fundamental and Applied Sciences, Universiti Teknologi PETRONAS, Seri Iskandar, Perak Darul Ridzuan, Malaysia;
Institute of Contaminant Management, Universiti Teknologi PETRONAS, Seri Iskandar, Perak Darul Ridzuan, Malaysia

Nur Natasha Bintang Mohd Jad Department of Chemical Engineering, Universiti Teknologi PETRONAS, Perak, Seri Iskandar, Malaysia

Nur Diyan Mohd Ridzuan Department of Fundamental and Applied Sciences, Universiti Teknologi PETRONAS, Perak, Seri Iskandar, Malaysia

Md. Mokarram Badsha University of Kuala Lumpur-Malaysian Institute Chemical & Bioengineering Technology (UniKL-MICET), Alor Gajah, Melaka, Malaysia

Mahaletchimi Murugan Department of Fundamental and Applied Sciences, Universiti Teknologi PETRONAS, Perak, Seri Iskandar, Malaysia

Suleiman Gani Musa Department of Fundamental and Applied Sciences, Universiti Teknologi PETRONAS, Seri Iskandar, Perak Darul Ridzuan, Malaysia;
Department of Chemistry, Al-Qalam University Katsina, Katsina, Nigeria

Anusha Nagaih PETRONAS Carigali Sendirian Berhad, Kuala Lumpur, Malaysia

Katsuya Nagase University of Yamanashi, Kofu Yamanashi, Japan;
Takahata Precision, Yamanashi, Japan

Mohammad Shakir Nasif Mechanical Engineering Department, Universiti Teknologi PETRONAS, Tronoh, Malaysia

Mohd Sofi Numin Department of Fundamental and Applied Sciences, Universiti Teknologi PETRONAS, Seri Iskandar, Perak, Malaysia

Noridah Binti Osman Center for Biofuel and Biochemical Research (CBBR), Universiti Teknologi PETRONAS, Bandar Seri Iskandar, Perak, Malaysia

Mohammad Nurizat Rahman Generation Unit, Generation and Environment, TNB Research, Kajang, Selangor, Malaysia

Raihan Mahirah Ramli Department of Chemical Engineering, Universiti Teknologi PETRONAS, Seri Iskandar, Perak, Malaysia

Mohamed Shuaib Mohamed Saheed Centre of Innovative Nanostructure and Nanodevices (COINN), Institute of Autonomous System, Universiti Teknologi PETRONAS, Seri Iskandar, Perak, Malaysia;
Department of Mechanical Engineering, Universiti Teknologi PETRONAS, Seri Iskandar, Perak, Malaysia

Nurul Tasnim Sahrin HICoE-Centre for Biofuel and Biochemical Research (CBBR), Institute of Self-Sustainable Building (ISB), Department of Fundamental and Applied Sciences, Universiti Teknologi PETRONAS, Perak Darul Ridzuan, Malaysia

Muhamad Shazarizul Haziq Mohd Samsuri Generation Unit, Generation and Environment, TNB Research, Kajang, Selangor, Malaysia

Akbar Abu Seman PETRONAS Research Sdn.Bhd., Kajang, Selangor, Malaysia

Tan Loo Sen PETRONAS Gas Berhad, Kuala Lumpur, Malaysia

Maizatul Shima Shaharun HICoE-Centre for Biofuel and Biochemical Research, Institute of Self-Sustainable Building, Department of Fundamental and Applied Sciences, Universiti Teknologi PETRONAS, Seri Iskandar, Perak Darul Ridzuan, Malaysia;
Institute of Contaminant Management, Centre for Contaminant Control & Utilization (CenCoU), Department of Fundamental and Applied Sciences, Universiti Teknologi PETRONAS, Seri Iskandar, Malaysia

Rashid Shamsuddin HICoE-Centre for Biofuel and Biochemical Research (CBBR), Institute of Self-Sustainable Building (ISB), Department of Chemical Engineering, Universiti Teknologi PETRONAS, Perak Darul Ridzuan, Malaysia

Ismail Shariff Generation Unit, Generation and Environment, TNB Research, Kajang, Selangor, Malaysia

Md. Sohrab Hossain HICoE-Centre for Biofuel and Biochemical Research, Institute of Self-Sustainable Building, Department of Fundamental and Applied Sciences, Universiti Teknologi PETRONAS, Seri Iskandar, Perak Darul Ridzuan, Malaysia;
University of Kuala Lumpur-Malaysian Institute Chemical & Bioengineering Technology (UniKL-MICET), Alor Gajah, Melaka, Malaysia

Nur Amirah Suhaimi Center of Contaminant Control and Utilization (CenCoU), Institute of Contaminant Management for Oil and Gas, Department of Fundamental and Applied Sciences, Universiti Teknologi PETRONAS, Seri Iskandar, Perak, Malaysia

Mohd Faisal Bin Taha Department of Fundamental and Applied Sciences, Universiti Teknologi PETRONAS, Bandar Seri Iskandar, Perak, Malaysia

Hiroki Takata Division of Applied Chemistry, Graduated School of Engineering, Sojo University, Kumamoto, Japan

Terath Kumar s/o Omporkas PETRONAS Carigali Sendirian Berhad, Kuala Lumpur, Malaysia

Ryuichi Tomoshige Division of Applied Chemistry, Gradated School of Engineering, Sojo University, Ikeda, Nishi-ku, Kumamoto, Japan

Farman Ullah Department of Fundamental and Applied Sciences, Universiti Teknologi PETRONAS, Seri Iskandar, Perak, Malaysia;
Centre of Innovative Nanostructure and Nanodevices (COINN), Institute of Autonomous System, Universiti Teknologi PETRONAS, Seri Iskandar, Perak, Malaysia;
Department of Physics, University of Science and Technology, Bannu, Khyber Pakhtunkhwa, Pakistan

Nor Hafizah Yasin PETRONAS Research Sdn.Bhd., Kajang, Selangor, Malaysia

Suzana Yusup Generation Unit, Generation and Environment, TNB Research, Kajang, Selangor, Malaysia

Noor Asmawati Mohd Zabidi Center of Contaminant Control and Utilization (CenCoU), Institute of Contaminant Management for Oil and Gas, Department of Fundamental and Applied Sciences, Universiti Teknologi PETRONAS, Seri Iskandar, Perak, Malaysia

Siti Nur Azella Zaine Department of Chemical Engineering, Universiti Teknologi PETRONAS, Perak, Seri Iskandar, Malaysia;

Centre of Innovative Nanostructure and Nanodevices (COINN), Institute of Autonomous System, Universiti Teknologi PETRONAS, Seri Iskandar, Perak, Malaysia

Siti Nur Azella Bt Zaine Centre of Innovative Nanostructures & Nanodevices (COINN), Universiti Teknologi PETRONAS, Bandar Seri Iskandar, Perak, Malaysia

Effect of Feedstock Composition on the Methanol Synthesis via the CO_2 Hydrogenation Process

Nor Hafizah Berahim, Noor Asmawati Mohd Zabidi, Raihan Mahirah Ramli, Nur Amirah Suhaimi, Akbar Abu Seman, and Nor Hafizah Yasin

Abstract Hydrogen has the potential to play a crucial part in the transition to CO_2-neutral industrial production, since its usage as an energy carrier and feedstock in numerous industrial process routes is promising. Chemical manufacturing, especially methanol, has considerable potential for hydrogen utilization. Methanol is an excellent hydrogen transporter and catalytic hydrogenation of CO_2 to methanol can address the issues of global warming and greenhouse gas emissions around the globe. In this regard, the present work highlights the effect of hydrogen utilization, combining with CO_2 as a feedstock for methanol synthesis using Cu/ZnO-based catalyst. The catalyst was synthesized, characterized and its performance was evaluated for hydrogenation of CO_2 at 250 °C, 32 bar, GHSV 6480 ml/g h with different H_2:CO_2 ratio of 3, 6.5 and 10. CO_2 conversion was increased sixfold when the H_2:CO_2 ratio was increased from 3 to 10. Nevertheless, methanol selectivity decreased from 18.12 to 6.17% when the H_2:CO_2 ratio increased from 6.5 to 10. Excess H_2 utilization resulted in increasing methyl formate formation.

N. H. Berahim (✉) · N. A. M. Zabidi · N. A. Suhaimi
Center of Contaminant Control and Utilization (CenCoU), Institute of Contaminant Management for Oil and Gas, Department of Fundamental and Applied Sciences, Universiti Teknologi PETRONAS, 32610 Seri Iskandar, Perak, Malaysia
e-mail: nor_19001638@utp.edu.my; hafizah.berahimjusoh@petronas.com

N. A. M. Zabidi
e-mail: noorasmawati_mzabidi@utp.edu.my

N. A. Suhaimi
e-mail: amirah.suhaimi@utp.edu.my

N. H. Berahim · A. A. Seman · N. H. Yasin
PETRONAS Research Sdn.Bhd., 43000 Kajang, Selangor, Malaysia
e-mail: akbar_seman@petronas.com

N. H. Yasin
e-mail: norhafizah.yasin@petronas.com

R. M. Ramli
Department of Chemical Engineering, Universiti Teknologi PETRONAS, 32610 Seri Iskandar, Perak, Malaysia
e-mail: raihan.ramli@utp.edu.my

M. B. Othman et al. (eds.), *Proceedings of the 1st International Conference of New Energy*, Springer Proceedings in Energy, https://doi.org/10.1007/978-981-99-0859-2_1

Keywords Hydrogen · CO_2 · Methanol · Cu/ZnO catalyst

1 Introduction

Alternative CO_2-based products may play a significant role in decoupling fossil resources from economic demands. Hydrogenation of CO_2 is one of the most easily implemented CO_2 conversion pathways for the production of chemicals [1]. Methanol has great potential as a hydrogen transporter molecule in chemical systems. The well-known industrial process for making methanol is a good example of how CO_2 and "green" hydrogen could be used to make a sustainable product [2]. In this context, the utilization of hydrogen as an intermediate material and energy source as a neutral-carbon chemical vector has been recognized [3–5]. Despite the fact that H_2 can be utilized as a raw material or fuel depending on the needs of the industry, the alternate approach would involve Carbon Capture and Utilization (CCU) technology, which would be a good option in managing the greenhouse gas CO_2 [6, 7].

The production of methanol (MeOH) depends heavily on hydrogen. Due to the fact that renewable hydrogen, which is often produced through electrolysis, is the most expensive part of the process, it has an impact on both the environment and economic performance. The process of carbon balance is improved by the addition of renewable hydrogen, but it requires careful planning to take into account the stochastic nature of renewable energy and its associated costs [8]. The synthesis of methanol from hydrogenation of CO_2 can be seen from two angles. CO_2 is transformed directly to methanol in Eq. 1, but there is also an indirect pathway via the reverse water–gas shift (RWGS), Eq. 2 and CO hydrogenation in Eq. 3.

$$CO_2 + 3H_2 \leftrightarrow CH_3OH + H_2O \quad H^\circ\, 298\,K = -49.5\,kJ/mol \tag{1}$$

$$CO_2 + H_2 \leftrightarrow CO + H_2O \quad H^\circ\, 298\,K = +41\,kJ/mol \tag{2}$$

$$CO + 2H_2 \leftrightarrow CH_3OH \quad H^\circ\, 298\,K = -90.5\,kJ/mol \tag{3}$$

The methanol synthesis process results in a close approximation of the thermodynamic equilibrium conversion. Any excess hydrogen is not used and remains unutilized in the reactor off-gases. Consequently, methanol synthesis is typically undertaken in a closed-loop system in which unreacted off-gases are recycled to the synthesis reactor [9]. Since hydrogen is the most expensive component in this scheme, an effective utilization strategy of the additional hydrogen could be key to making the process economically viable.

In this work, we report the utilization of H_2 at a different ratio in combination with captured CO_2 for methanol synthesis using tri-promoted Cu/ZnO supported on Al_2O_3 catalyst. The optimum H_2 usage to maximize methanol production via CO_2 hydrogenation using tri-promoted Cu/ZnO-based catalyst was highlighted.

2 Methodology

2.1 *Preparation of Catalyst Support*

Commercial Al_2O_3 was pretreated at 400 °C under Ar flow for 4 h with ramping 10 °C/min in a tube furnace.

2.2 *Supported Cu/ZnO-Based Catalyst with Addition of Promoter*

Cu/ZnO with fixed metal loading of 15 wt.% at a ratio of 7:3 and 0.09 wt.% of total promoters (Mn/Nb/Zr) have been synthesized. The amount of each precursor and promoter added were calculated based on the amount of catalyst mass prepared.

The metal precursors, copper nitrate trihydrate ($Cu(NO_3)_2{\cdot}3H_2O$) and zinc nitrate hexahydrate ($Zn(NO_3)_2{\cdot}6H_2O$), followed by promoters manganese (II) nitrate tetrahydrate ($Mn(NO_3)_2{\cdot}4H_2O$), ammonium niobate (V) oxalate hydrate ($C_4H_4NNbO_9{\cdot}xH_2O$) and zirconium (IV) oxynitrate hydrate ($ZrO(NO_3)_2{\cdot}H_2O$) were dissolved in deionized water to produce an aqueous solution. The solution was stirred for one hour. The prepared aqueous precursor solution was then added dropwise to the beaker containing the Al_2O_3 support. The pH of the mixture was kept at pH 7. The mixture was stirred for 24 h, filtered and washed with deionized water. The paste formed was dried in an oven at 120 °C for 12 h. The dried catalyst was then placed in a ceramic crucible and calcined at 350 °C for 4 h. The catalyst sample $Cu/ZnO/Mn/Nb/Zr/Al_2O_3$ was denoted as CZMNZA.

2.3 *Catalyst Characterization*

Morphologies of the samples were observed on the Hitachi SU8020 field emission scanning electron microscope (FESEM). Textural analysis has been carried out on a Micromeritics ASAP 2020 analyzer by determining the nitrogen adsorption/desorption isotherms at −196 °C. Prior to analysis, the samples were degassed at 350 °C (heating rate 10 °C/min) for 4 h. The Brunauer–Emmett–Teller (BET) specific surface area and pore volume are assessed from the adsorption data. The mean pore diameter is determined by applying the Barrett-Joyner-Halenda (BJH) model to the isotherm desorption branch [10]. X-ray diffraction (XRD) D2 Phaser from Bruker was employed, and PANanalytical High Score Plus software was used for phase identification. The XRD were measured at room temperature using 2θ (Bragg angle) ranging from 20° to 80°, with step size 0.024°/step and 1 s/step. The reduction behavior of the catalysts was studied using a Thermo Finnigan TPD/R/O 110 CE instrument equipped with a thermal conductivity detector (TCD). 40 mg of catalyst was placed in a quartz tube and pretreated with pure N_2 at 250 °C for 1 h at a 10 °C/min ramping rate to remove moisture and impurities. The analysis was then continued by switching to a 5% H_2/Ar (20 mL/min) gas flow to a maximum temperature of 950 °C with a 10 °C/min ramping rate and holding for an hour. The tail gas was then sent directly into the TCD, which calculated the amount of hydrogen consumed in the gas stream. Distinct reducible species in the catalyst were shown as peaks in the TPR spectrum. CO_2-TPD analysis was carried out using the same instrument as TPR to study the basic property of the catalyst. The calcined catalyst was pretreated at 250 °C for 50 min under He flow. CO_2 sorption was then resumed by flowing CO_2 at 10 mL/min for 30 min at 75 °C. The desorption of CO_2 was conducted by purging He gas with a flow rate of 20 mL/min through the sorbent bed and ramping the temperature from 40 to 950 °C at a rate 10 °C/min. The temperature was held at 110 °C for 20 min during the ramping sequence.

2.4 *Catalytic Performance Evaluation*

The catalyst performance for CO_2 hydrogenation to methanol was evaluated using a fixed-bed reactor (Effi microreactor, PID Eng & Tech). The catalyst was reduced *in-situ* at atmospheric pressure in H_2 for 6 h. Then, reactant gases containing H_2 and CO_2 with variation of feed ratio (3:1, 6.5:1 and 10:1) were fed into the reactor at fixed temperature of 250 °C, pressure of 32 bar and GHSV of 6480 ml/g.h. The gaseous products from the reactor were analyzed via an online gas chromatograph equipped with TCD and FID detectors. The catalytic performance was evaluated based on the CO_2 conversion and product selectivity calculated as per Eqs. 4 and 5.

$$\text{CO}_2\ \text{Conversion}\ (\%) = \frac{\text{moles of CO}_{2\text{in}} - \text{moles of CO}_{2\ \text{out}}}{\text{moles of CO}_{2\ \text{in}}} \times 100 \tag{4}$$

$$\text{MeOH Selectivity}\ (\%) = \frac{\text{moles of MeOH produced}}{\text{total moles of products}} \times 100 \tag{5}$$

3 Results and Discussion

3.1 *Morphology of Catalyst*

The morphology of the Al_2O_3 bare support and impregnated catalyst CZMNZA were analyzed using FESEM, and the resulting images are shown in Fig. 1. The commercial Al_2O_3 support and CZMNZA exhibited irregular morphologies. The metal oxide nanoparticles of CZMNZA were unevenly distributed on the Al_2O_3 surface. Some of the promoters are not detected in the sample due to low amount

Fig. 1 FESEM images of **a** Al_2O_3 **b** CZMNZA and EDX mapping of **c** CZMNZA

of promoters added in the formulation and sample inhomogeneity as shown by the EDX mapping in Fig. 1c.

3.2 *Textural Properties*

The textural properties of the samples are shown in Table 1. After metal oxide loading, the surface area (S_{BET}) and pore volume (V_P) of Al_2O_3 bare support are reduced to some extent, indicating that the active phase was distributed onto the support and filled the pores [11]. In contrast, the pore diameter increased with the addition of metal oxide. It is suggested that the particles will selectively close the smaller pores during the loading. As a result, pore volume was lost, but the average pore width grew since the larger pore was the only one still available for physisorption.

Table 1 Textural properties of the samples

Sample	S_{BET} (m^2/g)	V_P (cm^3/g)	D_{BJH} (nm)
Al_2O_3	134.51	0.26	5.29
CZMNZA	109.33	0.22	5.65

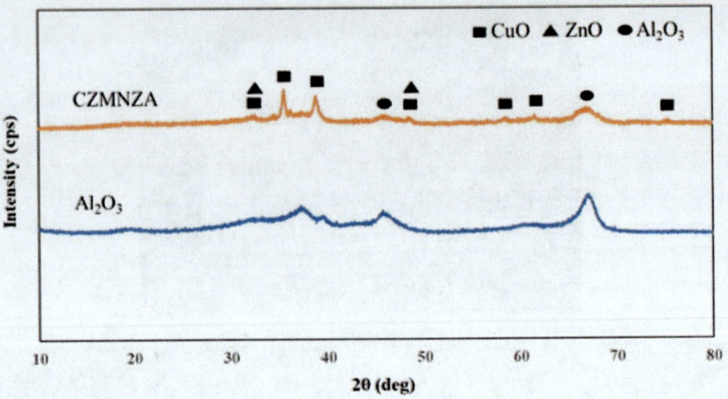

Fig. 2 XRD diffraction patterns for Al_2O_3 and CZMNZA

3.3 *Phase Analysis*

Figure 2 demonstrates the XRD patterns of Al_2O_3 bare support and its impregnated catalysts, CZMNZA. Two significant peaks at 45.90° and 67.18° can be identified, indicating the presence of gamma aluminum oxide (γ-Al_2O_3) [12] and the peaks are evident in the diffraction peaks of the CZMNZA, which indicate the existence of alumina as a catalyst support. Tenorite CuO peaks were found in CZMNZA sample at 2θ 32.5°, 35.5°, 38.8°, 48.8°, 58.3°, 61.5°, and 72.5° [12]. On the other hand, zincite, ZnO diffraction peaks were observed at 32.5° and 48.8°, although the peak may overlap with CuO peaks [13, 14].

3.4 *Reducibility and Basicity Studies*

The ability of a catalyst to absorb hydrogen as temperature increases can provide information on the appropriate temperature for the catalyst to be reduced into its active phase. The H_2-TPR profile of the CZMNZA catalyst is depicted in Fig. 3. The α peak is associated with the reduction of highly dispersed Cu species, whereas the β peak is related to bulk-like Cu species [15, 16]. The reduction for the CZMNZA occurred below 400 °C with α peak temperature at 214 °C and β peak at 269 °C. Based on the previous study related to catalyst activation condition, the catalyst was reduced at 243 °C for six hours prior to CO_2 hydrogenation process.

Since CO_2 gas is acidic enough to reach all the basic sites, it was used in the experiment to identify the density and strength of the catalyst basic sites. The CO_2 TPD profile of CZMNZA is shown in Fig. 4 and the quantitative data was summarized in Table 2. Depending on their temperature, basic sites can be classified as weak, α (<200 °C), medium, β (200–550 °C), or strong, γ (>500 °C) [15, 17]. The α and β basic sites are the anticipated active sites for the CO_2 hydrogenation process given that the reaction temperature is kept below 400 °C.

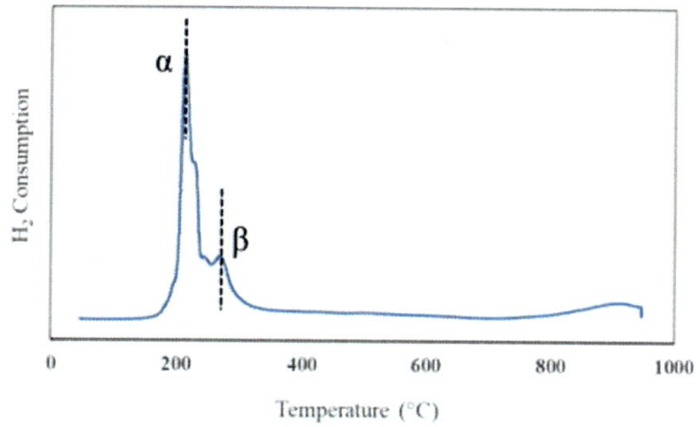

Fig. 3 H_2-TPR profile for CZMNZA

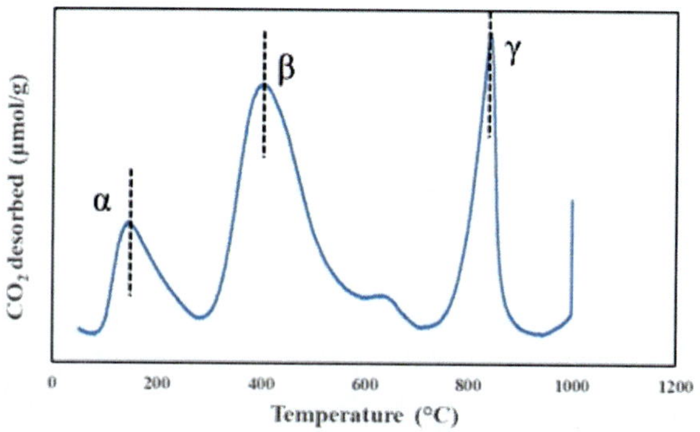

Fig. 4 CO_2 TPD profile of CZMNZA

Table 2 Quantitative data for CO_2 TPD of CZMNZA

Sample	T (°C)	Basic site (μmol/g)	T (°C)	Basic site (μmol/g)	T (°C)	Basic site (μmol/g)	Total basic site (μmol/g)	α + β fraction
CZMNZA	137	34	382	97	897	88	219	0.59

3.5 *Catalytic Activity*

The effect of H_2 utilization at different ratios was tested in a CO_2 hydrogenation reaction for methanol synthesis. Reaction temperature, pressure and space velocity were fixed at 250 °C, 32 bar and 6480 ml/g.h, respectively. Table 3 shows the catalytic performance in terms of CO_2 conversion and product's selectivity. Methanol (MeOH) and methyl formate (MF) are the products that have been quantified in the CO_2 hydrogenation reaction. Increasing the amount of H_2 in the feedstock enhanced the CO_2 conversion, in accordance with Le-Chatelier's principle. Nevertheless, methanol selectivity decreased from 18.12 to 6.57% and methyl formate increased from 81.67 to 93.43% when the ratio of H_2 was increased from 6.5 to 10. The methyl formate could be formed via the following competition reaction:

Table 3 Catalytic activity of CZMNZA

H_2: CO_2	CO_2 conversion (%)	Selectivity (%)	
		MeOH	MF
3	2.00	14.80	84.85
6.5	6.35	18.12	81.67
10	12.02	6.57	93.43

$$2CO + 2H_2 \rightarrow HCOOCH_3 \quad (6)$$

The presence of excess hydrogen could increase the rate of the RWGS reaction which produces CO (Eq. 2). The generated CO could subsequently react with the excess hydrogen leading to higher amount of methyl formate formation with increasing fraction of hydrogen in the feed gas stream. The utilization of H_2 at different ratios significantly affected both CO_2 conversion and product's selectivity.

4 Conclusion

Effect of H_2 utilization in a CO_2 hydrogenation reaction in the presence of tri-promoted Cu/ZnO/Al_2O_3 catalyst was investigated. The value of CO_2 conversion increased when H_2: CO_2 ratio was increased. However, further increase in the H_2:CO_2 ratio from 6.5 to 10 resulted in a threefold decrease in methanol selectivity and increased the methyl formate formation. Operating parameters perhaps need to be optimized which could lead to more efficient utilization of H_2 for the optimum catalytic performance in the methanol synthesis reaction.

Acknowledgements The authors acknowledge the financial support provided by PETRONAS Research Sdn.Bhd. and Universiti Teknologi PETRONAS (UTP) under MRA Research grant (cost center 015MDO-031).

References

1. J. Fernández-González, M. Rumayor, A. Domínguez-Ramos, A. Irabien, Hydrogen utilization in the sustainable manufacture of CO_2-based methanol. Ind. Eng. Chem. Res. **61**(18), 6163–6172 (2022)
2. M. Behrens, Chemical hydrogen storage by methanol: challenges for the catalytic methanol synthesis from CO_2. Recycl. Catal. **2**(1), 78–86 (2016)
3. Cembureau, Cementing the European Green Deal. 101(2003), 16 (2016)
4. Eurofer, Low carbon roadmap: pathways to a CO_2-neutral European steel industry. The European Steel Association (2019)
5. V.M. Maestre, A. Ortiz, I. Ortiz, Challenges and prospects of renewable hydrogen-based strategies for full decarbonization of stationary power applications. Renew. Sustain. Energy Rev. **152**, 111628 (2021)

6. IEA, Global energy review: CO_2 emissions in 2020 (2021)
7. UNFCCC, The Paris agreement. United Nations (2016)
8. M. Bampaou, A.S. Kyriakides, K. Panopoulos, P. Seferlis, S. Voutetakis, Modelling of methanol synthesis: improving hydrogen utilisation. Chem. Eng. Trans. **88**, 931–936 (2021)
9. I. Løvik, Modelling, estimation and optimization of the methanol synthesis with catalyst deactivation. Ph.D. Thesis, Norwegian University of Science and Technology (2001)
10. E.P. Barrett, L.G. Joyner, P.P. Halenda, The determination of pore volume and area distributions in porous substances. I. computations from nitrogen isotherms. J. Am. Chem. Soc. **73**(1), 373–380 (1951)
11. S.F.H. Tasfy, N.A.M. Zabidi, M.S. Shaharun, D. Subbarao, The role of support morphology on the performance of Cu/ZnO-catalyst for hydrogenation of CO_2 to methanol. AIP Conf. Proc. **1669**(1) (2015)
12. N.S.A. Halim, Synthesis, characterization and performance of copper-zinc oxide with mixed promoter catalyst for CO_2 hydrogenation to methanol. Master Thesis. Universiti Teknologi Petronas (2019)
13. C. Huang, S. Chen, X. Fei, D. Liu, Y. Zhang, Catalytic hydrogenation of CO_2 to methanol: study of synergistic effect on adsorption properties of CO_2 and H_2 in $CuO/ZnO/ZrO_2$ system. Catalysts **5**(4), 1846–1861 (2015)
14. Z. Li, S. Yan, H. Fan, Enhancement of stability and activity of $Cu/ZnO/Al_2O_3$ catalysts by microwave irradiation for liquid phase methanol synthesis. Fuel **106**, 178–186 (2013)
15. O. Tursunov, L. Kustov, Z. Tilyabaev, Methanol synthesis from the catalytic hydrogenation of CO_2 over CuO–ZnO supported on aluminum and silicon oxides. J. Taiwan Inst. Chem. Eng. **78**, 416–422 (2017)
16. P. Lackner, Z. Zou, S. Mayr, S. Diebold, M. Schmid, Using photoelectron spectroscopy to observe oxygen spillover to zirconia. Phys. Chem. Chem. Phys. **21**(32), 17613–17620 (2019)
17. V.D.B.C. Dasireddy, B. Likozar, The role of copper oxidation state in $Cu/ZnO/Al_2O_3$ catalysts in CO_2 hydrogenation and methanol productivity. Renew. Energy **140**, 452–460 (2019)

Inhibition of Ammonia Emission by Buffer Solution in Ammonia Borane Hydrolysis

Hiroki Takata and Hitoshi Inokawa

Abstract Ammonia borane (AB) is a promising hydrogen storage material for fuel cells because of its high hydrogen density and its ability to produce hydrogen through catalytic hydrolysis reaction at room temperature, although ammonia is also generated during the hydrolysis, resulting in damages to fuel cells. In this study, we investigated the AB hydrolysis reaction in phosphate buffer solution, acetate buffer solution, and citrate buffer solution, in order to reduce ammonia release. As a result, all buffer solutions brought a significant reduction of ammonia concentration in released hydrogen compared to pure AB solution without pH buffer. Mass spectroscopy in-situ analysis revealed that ammonia concentration in hydrogen released from the pure AB solution reached a high level after starting the reaction and gradually increased over the reaction time, whereas that from phosphate buffer solution was kept at a low level during the hydrolysis reaction. Attenuated total reflectance (ATR) infrared (IR) in-situ spectroscopy of AB solution with phosphate buffer demonstrated that $H_2PO_4^-$ released H^+ and formed HPO_4^{2-}. Therefore, it was concluded that the addition of pH buffer into AB solution drastically suppress ammonia release during the catalytic hydrolysis reaction, and phosphate buffer is the most promising.

Keywords Hydrogen · Borohydride · Catalyst · Fuel cell

1 Introduction

Due to energy and environmental issues, research and development of hydrogen storage and transportation technologies have been actively conducted. Ammonia Borane (AB) has the highest hydrogen density among amine boranes at 19.6 wt% and is stable at room temperature [1]. AB can be completely dehydrogenated by thermal decomposition, which requires heating above 500 °C [2]. Hydrogen can be released from AB even at room temperature by a hydrolysis reaction (Eq. 1). The

H. Takata · H. Inokawa (✉)
Division of Applied Chemistry, Graduated School of Engineering, Sojo University, 4-22-1 Ikeda, Nishi-Ku, Kumamoto, Japan
e-mail: inokawa@nano.sojo-u.ac.jp

M. B. Othman et al. (eds.), *Proceedings of the 1st International Conference of New Energy*, Springer Proceedings in Energy, https://doi.org/10.1007/978-981-99-0859-2_2

hydrolysis of AB can be initiated by an addition of metallic catalysts such as Ni [3], Co [4], Ru [5], and Pt [6].

Regarding the hydrolysis, a half of the generated hydrogen is derived from water and the other half are from BH_3 group of AB (Eq. 1).

$$NH_3BH_3 + 3H_2O \xrightarrow{cat} 3H_2 + B(OH)_3 + NH_3 \quad (1)$$

The AB hydrolysis reaction results in generation of ammonia, which can be partially dissolved in water but is also released with hydrogen. The amount of the released ammonia depends on the experimental conditions, for example concentration of AB, reaction temperature, and so on. Because ammonia causes a significant damage to fuel cells, the release of ammonia is an important issue and requires the use of a filter between the fuel cell and the hydrogen generator, making the overall system more complex and increasing costs. This makes the superiority of sodium borohydride over hydrolysis problematic [7]. Thus, the generation of ammonia is an issue to be solved for the practical use of AB, but no effective solution has been proposed yet.

In order to solve this, we investigated the reduction of ammonia released from the reactant solution to the gas phase during the hydrolysis reaction by adding a pH buffer into AB solution, rather than removing ammonia from generated hydrogen by ammonia adsorbents. In the AB solution, the pH buffer can produce proton, which converts the ammonia into ammonium ion and captures them in the aqueous solution. In this study, we report the effects of different types of buffers, which are phosphate buffer, citrate buffer, and acetate buffer, on the hydrogen and ammonia release behavior during the AB hydrolysis reaction.

2 Experimental Methods

2.1 *Catalyst Preparations*

An incipient wetness impregnation method was used to support 1 wt% Pt catalyst on activated carbon (AC, Nacalai Tesque, INC.). AC powder was added to chloroplatinic acid (Sigma-Aldrich) aqueous solution, and the water was removed by a rotary evaporator. After the obtained powder was reduced with $NaBH_4$ aqueous solution, the catalyst was collected via filtration, washing, and drying in an oven at 60 °C.

2.2 *Ammonia Emission Control Test*

The concentration of ammonia released from AB solution was typically evaluated by the following procedure. pH buffer solution was prepared by dissolving a proton donating material, which was potassium dihydrogen phosphate (Fujifilm Wako Pure Chemical Corporation), acetic acid (Kanto Chemical CO., INC.), or citric acid (Kanto Chemical CO., INC.), into 50 mL of deionized water with its counterpart salt, which was disodium hydrogen phosphate (Fujifilm Wako Pure Chemical Corporation), sodium acetate trihydrate (Fujifilm Wako Pure Chemical Corporation), or tripotassium citrate monohydrate (Fujifilm Wako Pure Chemical Corporation), respectively. The amount of the proton donors was equivalent to mols of AB. Initial pH of the buffer solutions was adjusted at approximately 6.5 by dissolving the salts. Then, 100 mg of AB (Tokyo Chemical Industry CO., LTD.) was added to the pH buffer solutions. After AB was dissolved completely in the buffer solutions, AB hydrolysis reaction took place by adding 50 mg of 1 wt% Pt/AC catalyst. The produced gas was collected in a tedlar bag, and ammonia concentration was analyzed by a gas detector tube after finishing the hydrolysis reaction. The amount of the produced gas was monitored by a water displacement system. The tedlar bag was placed in a flask filled with water, and the generated gas expanded the bag and forced out the water from inside of the flask to another container on a balance. In order to compare ammonia emission between AB solutions with or without pH buffer materials, the catalytic hydrolysis reaction in pure AB aqueous solutions was performed, and ammonia concentration was measured by the same method mentioned above. Hydrolysis reactions with various AB and phosphate buffer concentrations were also performed by decreasing the amount of water in order to evaluate ammonia release behavior from the condensed reactant solutions.

For an in-situ analysis of gaseous products, AB aqueous solution with/without phosphate buffer was placed at the side of a forked(two-way) test tube, and 1 wt% Pt/AC catalyst was placed at another side of the tube, followed by continuous Ar flow connected to a Mass Spectrum (MS) analysis. AB hydrolysis reaction was started by transferring the reactant solution to the catalyst and the gas was analyzed over the reaction by the MS.

2.3 *Analysis of the Mechanism of Ammonia Release Inhibition*

To investigate the effect of proton donors, reactant solutions were prepared with varying H_2KPO_4/AB molar ratio from 0.54 to 1.20, and the concentration of the released ammonia was measured after termination of AB hydrolysis reaction.

Neutralization titration was performed by adding 0.325 M NaOH solution into each buffer solution to investigate the buffering capacity.

To confirm the proton donation from the buffer solution during the AB hydrolysis reaction, AB solution containing phosphate buffer was placed with catalyst on a ZnSe-crystal of Attenuated total reflectance (ATR) infrared (IR) spectroscopy, and in-situ ATR-IR spectra of the hydrolysis reaction were recorded every 5 min.

Table 1 Emitted ammonia by AB hydrolysis with/without pH buffer

	No buffer	Phosphate buffer	Citrate buffer	Acetate buffer
NH_3 Emission(ppm)	100	3	4	6

3 Results and Discussion

3.1 Emission Behavior of Ammonia

Ammonia concentrations released from AB hydrolysis with/without pH buffers were organized in Table 1. While the amount of ammonia released from 0.065 M AB solution without buffer was 100 ppm, the amount of ammonia released with any of the buffer solutions did not exceed 6 ppm, indicating that the use of buffer solutions in the reaction field of AB hydrolysis is very effective for inhibiting ammonia release. Regarding hydrogen production behavior, the acetate buffer solution and citrate buffer solution brought lower hydrogen production rate than pure AB solution, whereas phosphate buffer solution showed similar hydrogen production behavior as pure AB solution (Fig. 1). Therefore, the phosphate buffer is the most appropriate of the three to be added into AB solution. The results of MS in-situ analysis are shown in Fig. 2. Ammonia is generally monitored with spectra of $m/z = 17$, 16, and 15, which are NH_3, and fragments of NH_2 and NH, respectively. Water (H_2O, $m/z = 18$), however, also affect spectra of $m/z = 17$ and 16 because of fragments of OH and O, respectively. Therefore, only the spectrum of $m/z = 15$ is appropriate to analyze the ammonia emission behavior. It is demonstrated that the release of ammonia increased as the decomposition of AB proceeded without buffer solution, but the release of ammonia was maintained at a low concentration during the reaction by adding phosphate buffer (Fig. 2). Ammonia emission from 0.32 M AB solution without buffer is about 1000 ppm, suggesting the increase of ammonia emission as increasing AB concentration. In case of condensed AB solution, which were 0.32 or 2.57 M, ammonia release was significantly suppressed by adding phosphate buffer as shown in Table 2.

3.2 The Mechanism of Ammonia Emission Control

In experiments which varied H_2KPO_4/AB molar ratio from 0.54 to 1.20 equivalents, a negative correlation was observed between the amount of proton donor (H_2KPO_4) and the amount of ammonia, as shown in Fig. 3. This suggests that the buffer supplies the protons required for the protonation of ammonia in the reactant solution at H_2KPO_4/AB molar ratio >1.

In fact, the buffering capacity of each buffer solution was measured by titration of NaOH solution as shown in Fig. 4. An equivalent amount of NaOH to

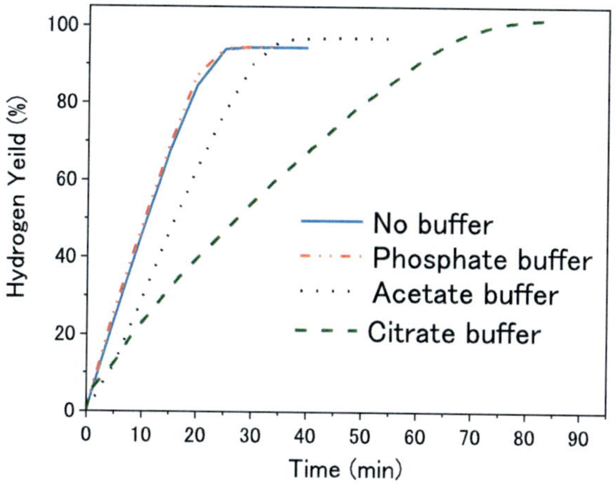

Fig. 1 Hydrogen generation property without/with phosphate buffer, acetate buffer, and citrate buffer

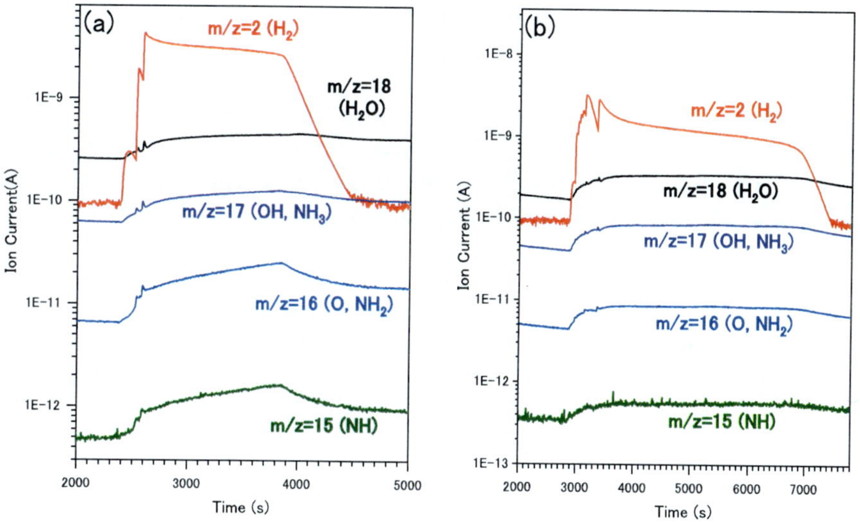

Fig. 2 In-situ MS spectra during AB hydrolysis reaction at 0.32 mol/L (**a**) without and (**b**) with phosphate buffer

Table 2 Ammonia emission at different AB concentrations with/without phosphate buffer

Concentration of AB (mol/L)	0.065	0.32	2.57
Amount of solvent (mL)	50	10	1.26
Ammonia emission with phosphate buffer (ppm)	2	5	2
Ammonia emission without pH buffer (ppm)	100	980	> 1000 *Beyond the upper limit

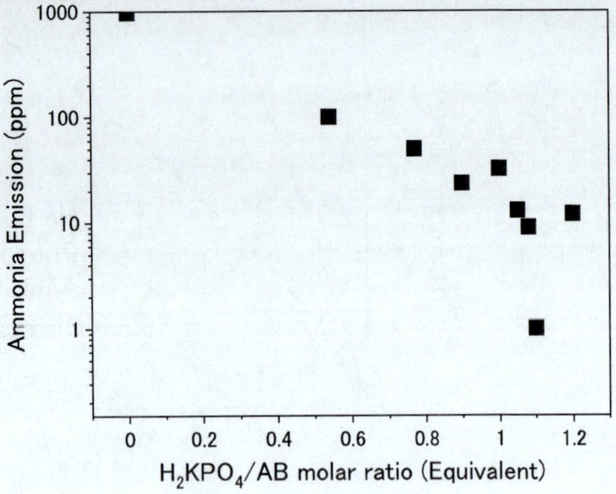

Fig. 3 Dependence of ammonia emission on H_2KPO_4/AB molar ratio (Equivalent) at initial pH 7.5 via AB hydrolysis reaction with phosphate buffer

ammonia which can be generated from the AB hydrolysis reaction (Eq. 1) under the same condition was about 3.2 mmol, and shown with a black vertical line in Fig. 4. It was confirmed that the pH of each solution can be maintained below 9.0 even if all ammonia molecules generated by the hydrolysis reaction dissolved in the buffer solutions. This suggests that the buffer solutions have enough capacity to capture ammonia molecules generated via the AB hydrolysis reaction. Figure 5 shows the in-situ ATR-IR spectra of AB hydrolysis reaction in phosphate buffer solution. According to Yang et al., a peak assigned to symmetric P-O stretch vibration of HPO_4^{2-} appeared at lower wavenumber than that of $H_2PO_4^-$ [8]. In Fig. 5, a peak assigned to the symmetric P-O stretch vibration appeared at 1065 cm^{-1} at the initiation of the AB hydrolysis reaction, and shifted to lower wavenumber with the progress of the reaction, suggesting that $H_2PO_4^-$ was consumed and transformed to HPO_4^{2-}. This is consistent with the concept of this study, which proton donor ($H_2PO_4^-$) of pH buffer protonates the generated ammonia and captures it as ammonium ion in the reaction solution.

4 Conclusions

pH buffer materials, which were phosphate buffer, citrate buffer, and acetate buffer, were added into AB solution in order to suppress ammonia release from the reactant solution. All pH buffers significantly reduced ammonia concentration in the released hydrogen. However, acetate buffer and citrate buffer decreased hydrogen generation rate compared to pure AB solution. Phosphate buffer is the most promising of the three because it showed similar hydrogen generation rate to pure AB solution. The effect to suppress the ammonia emission depended on M/AB molar ratio: Where M is proton donor of the pH buffer, for example H_2KPO_4 of phosphate buffer. The ammonia suppression was maximized at M/AB > 1.

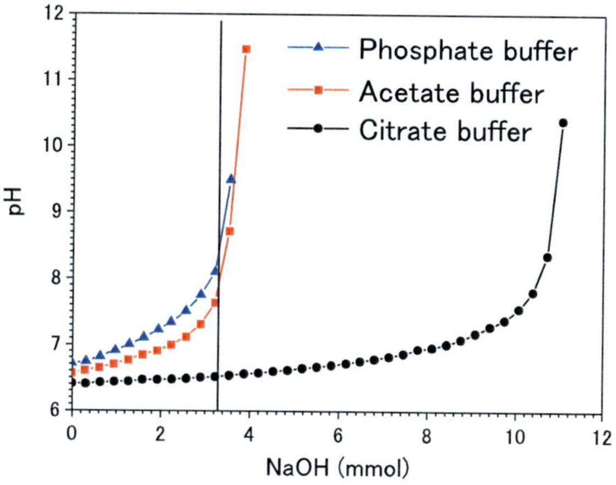

Fig. 4 Determination of buffering capacity in each buffer solution

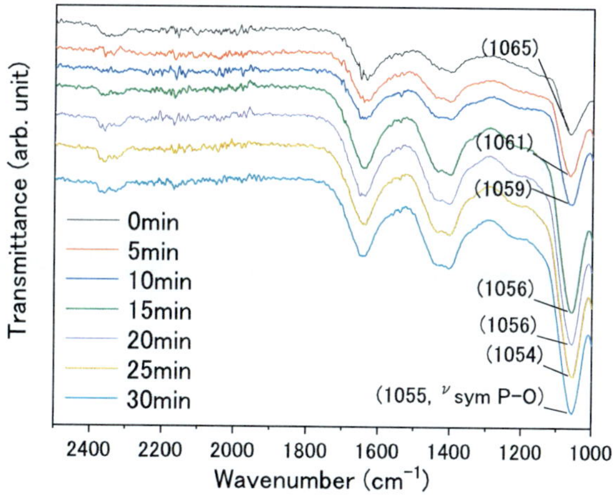

Fig. 5 In-situ ATR-IR of AB solution with phosphate buffer and catalyst recorded at every 5 min after starting the hydrolysis reaction

In-situ MS analysis of gas species during AB hydrolysis reaction revealed that ammonia concentration in hydrogen generated from pure AB solution was high and kept increasing during the reaction, whereas that from AB solution containing phosphate buffer was low and constant.

Because ammonia causes significant damage to fuel cells, ammonia concentration in hydrogen for fuel cells must be less than 0.1 ppm, according to ISO/TS14687-2. There is a huge difficulty for hydrogen generator using conventional AB hydrolysis to achieve the quite low ammonia concentration by using ammonia filter. The filter unit can cause the increasing in whole size of the H_2 generation unit and pressure loss. The results of this study contribute to a compact filter unit with long life time. In addition, even if the filter unit lost the ammonia capturing property, fuel cells could avoid being exposed to a gas containing ammonia with quite high concentration.

Acknowledgements A part of results in this study were obtained as a result of a project (JPNP20003) funded by the New Energy and Industrial Technology Development Organization (NEDO).

References

1. U.B. Demirci, Ammonia borane, a material with exceptional properties for chemical hydrogen storage. Int. J. Hydrogen energy **42**, 9978–10013 (2017)
2. N. Mohajeri, A. T-Raissi, O. Adebiyi, Hydrolytic cleavage of ammonia-borane complex for hydrogen production, J. Power Sources **167** 482–485 (2007)
3. T. Umegaki, Q. Xu, Y. Kojima, In situ synthesized spherical nickel-silica composite particles for hydrolytic dehydrogenation of ammonia borane. J. Alloys Compd. **580**, S313–S316 (2013)
4. M. Rakap, Hydrogen generation from the hydrolytic dehydrogenation of ammonia borane using electrolessly deposited cobalt-phosphorus as reusable and cost-effective catalyst. J. Power Sources **265**, 50–56 (2014)
5. S. Akbayrak, S. Tanyıldızı, I. Morkan, S. Ozkar, Ruthenium (0) nanoparticles supported on nanotitania as highly active and reusable catalyst in hydrogen generation from the hydrolysis of ammonia borane. Int. J. Hydrogen Energy **39**, 9628–9637 (2014)
6. W. Chen, J. Ji, X. Duan, G. Qian, P. Li, X. Zhou et al., Unique reactivity in Pt/CNT catalyzed hydrolytic dehydrogenation of ammonia borane. Chem. Commun. **50**, 2142–2144 (2014)
7. U.B. Demirci, Ammonia borane: an extensively studied, though not yet implemented, hydrogen carrier. Energies **13**, 3071 (2020)
8. K. Yang, R. Kas, W.A. Smith, In Situ Infrared Spectroscopy Reveals Persistent Alkalinity near Electrode Surfaces during CO_2 Electroreduction. J. Am. Chem. Soc. **141**, 15891–15900 (2019)

Cold Startup of a PEFC Studied by *Operando* Visualization of Ice and Oxygen Partial Pressure

Katsuya Nagase and Junji Inukai

Abstract During operation of a polymer electrolyte fuel cell (PEFC) at a subzero temperature, the water generated by the oxygen reduction reaction at the cathode tends to be frozen. During the cold startup test at −10 °C of a PEFC with five straight channels, the visualization of p_{O2} on the surface of gas diffusion layer was carried out using an oxygen sensitive dye. During the cooling down of the cell filled with dry nitrogen from 40.5 to −10 °C, the cell resistance increased. The dried mixture gas (N_2/O_2) supplied at −10 °C increased the cell resistance. During the cold startup, the current density was increased from 2 to 30 mA cm^{-2}. Before the voltage drop, oxygen was found to be consumed near the inlet. The central three channels generated more power than the side two channels.

Keywords PEFC · Subzero startup · Visualization of oxygen distribution

1 Introduction

Polymer electrolyte fuel cells (PEFCs) are widely used for residential co-generation systems and automobiles. For wider commercialization of these devices, especially for automobiles, stable performance over a wide range of cell temperature, including the subfreezing condition, is one of the most important issues. During the operation of a PEFC at a subzero temperature, the water generated by the oxygen reduction reaction in the cathode tends to be frozen. It was reported that the freezing of generated water can lead to increased ohmic and mass transport losses, due to delamination of the cathode catalyst layer (CL) and gas diffusion layer (GDL), as well as the collapse and densification of the cathode CL [1–4]. Electrochemical impedance spectroscopy has traditionally been used to qualify and quantify proton conductivity

K. Nagase · J. Inukai (✉)
University of Yamanashi, 4-3-11 Takeda, Kofu Yamanashi 400-8510, Japan
e-mail: jinukai@yamanashi.ac.jp

K. Nagase
Takahata Precision, 390 Maemada, Sakaigawa-cho, Fuefuki, Yamanashi 406-0843, Japan

M. B. Othman et al. (eds.), *Proceedings of the 1st International Conference of New Energy*, Springer Proceedings in Energy, https://doi.org/10.1007/978-981-99-0859-2_3

of the membrane [1]. The morphology of the cathode CL and the GDL was examined. Cracks and catalyst domain segregation in fully hydrated MEAs are reported after freeze–thaw cycles between 20 and −30 °C [5]. The collapses were led by the microstructural and chemical changes in degraded MEAs, which were investigated by transmission electron microscopy, scanning electron microscopy, and X-ray diffraction [5].

Better solutions for water management in the MEA are apparently needed. Various approaches have been proposed to alleviate the problems caused by the resulting ice plugging the gas diffusion networks [3, 6–14]. Experimental approaches have been also reported for the improvement of PEFC cold-start performance [8, 11, 12, 15, 16]. Uchida and coworkers reported the effects of the pore diameter of the GDL on PEFC performance at subfreezing temperatures [3]. The smaller pore size of GDL showed favorable cold-start performance in the ramped current mode at −5 °C, but the difference hardly affected the water removal performance below −10 °C. The reduced conductivity at subzero temperatures has been studied by the analysis of the water state in membranes measured via the differential scanning calorimetry [2, 17, 18]. It was reported that the conductivity of the membrane is affected by the free frozen water, non-freezing water, and shielding effect of ice in the membrane on proton transport.

Visualization approaches have been used to provide insight into water formation within the MEAs and GDLs under subzero conditions, thereby attempting to draw relationships between cold-start performance and appearance of water [2, 10, 19, 20]. A cryo-field emission scanning electron microscopy (FE-SEM) analysis method has been designed to visualize in situ the freezing process of product water [19], and the ice distribution in the CL and GDL under subzero operation was characterized with nanometer scale resolution. The result showed that the isothermal cold operation was not possible at the subfreezing temperature due to the pores being nearly filled by generated water. Neutron imaging was also used as a non-invasive visualization tool adapted to track the evolution of condensed phases of water (liquid, ice, membrane water content) in operating PEFCs [2]. When purging the cell before cooling, the super-cooled water emerging on the GDL surface during the cold start was observed. After a few minutes of operation, water freezing was localized at the MEA/GDL interface and was correlated with the voltage drop. These experiments showed the locations of ice and the presence of super-cooled water. However, the visualization of gasses under subzero conditions has not been carried out, which should directly show the influence of liquid/ice water on the gas diffusivity reflecting the cell performance. The measurements of the gas diffusion to the CL are required, as well as the gas consumption, during the power generation under the subzero temperatures.

We have visualized the oxygen partial pressure (p_{O2}) and ice water simultaneously inside an operating PEFC at subfreezing temperature. During the cold startup of a PEFC, the power generation reaction proceeded inhomogeneously with the oxygen consumption at specific locations on the catalyst layer.

2 Experimental

The preparation of MEA was described elsewhere [20, 21]. In this experiment, 24BCH of SGL was used for a GDL. The cell temperature for the startup tests was set at −10 °C. Images were captured every 10 s during the cell operation. The spatial resolution was 120 μm. Figure 1 shows the optical system for the cold startup. For controlling the cell temperature, an environmental chamber with a window made of transparent quartz was used. Dried air was introduced into the chamber for avoiding icing.

Before the cold startup test, the cell temperature was set at 40.5 °C. Humidified nitrogen (97% RH) was supplied at 0.100 $dm^3\ min^{-1}$ to both the anode and cathode for 3 h. Then, dry nitrogen was supplied to the anode and cathode until the cell resistance became as large as 300 mΩ, when nitrogen purging and cell heating were stopped. Then, the cell filled with dry nitrogen was cooled to −10 °C. For the p_{O2} visualization, calibration curves were subsequently obtained at −10 °C using the same cell. For obtaining calibration curves at each pixel of a CCD camera, the cathode was supplied with a mixture of oxygen and nitrogen at 0.400 $dm^3\ min^{-1}$ for purging at the ratios from 0 to 25% for the oxygen concentration, while dry nitrogen was continuously supplied in the anode at 0.100 $dm^3\ min^{-1}$. In order to minimize the pressure loss [21], the flow rate of the cathode gas was lowered to 0.050 $dm^3\ min^{-1}$ during the emission measurements for the calibration. After obtaining the calibration plots, dry air (0.400 $dm^3\ min^{-1}$) was supplied to the cathode followed by dry nitrogen. Before power generation, dried hydrogen and air (0.100 $dm^3\ min^{-1}$) were supplied to the anode and cathode, respectively. After the open-circuit voltage reached 1 V, the flow rate of air decreased to 0.030 $dm^3\ min^{-1}$ and the cold startup test began at the current density of 2 mA cm^{-2}. Whenever the cell voltage reached 0.6 V, the current density was increased to 6, 10, 15, 20, and 30 mA cm^{-2}. During the startup tests, the emission

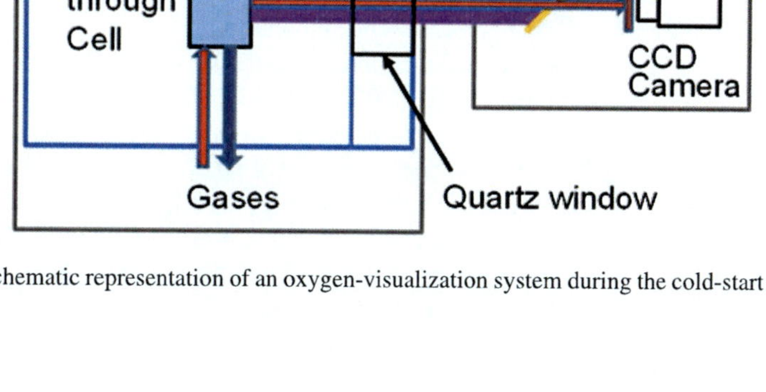

Fig. 1 Schematic representation of an oxygen-visualization system during the cold-start operation

images from the dye film coated on the GDL surface were continuously captured. A high frequency resistance of the cell was also measured with an AC milliohmmeter, (Model 3566 Tsuruga Electric Corporation) at 1 kHz.

3 Results and Discussion

Figure 2a shows the images of the steady-state distributions of p_{O2} on the GDL surface at a current density of 0.6 A cm^{-2} (except the oxygen utilization (U_{O2}) of 0%) obtained at the cathode during the cell operation at a cell temperature of 80 °C with humidified air and hydrogen (80% RH) [20]. The hydrogen-flow rate was 0.200 dm^3 min^{-1}. U_{O2} values were 0, 15, and 30% at the air-flow rates of 0.332, 0.111, and 0.055 dm^3 min^{-1}, respectively. The color of the images corresponds to p_{O2}. The p_{O2} values on the central channel, Channel 3, were plotted along the channel length in Fig. 2b. We have reported that the p_{O2} values are different across the channel width [23], so the p_{O2} values in the middle of Channel 3 are employed in Fig. 2b. At the open-circuit voltage with no power generation (U_{O2} = 0%), p_{O2} was 13 kPa in all channels (left image in Fig. 2a, b). The pressure loss was very small compared with that in the serpentine channels [20, 21, 23, 24]. When the cell was operated at U_{O2} = 15% (0.111 dm^3 min^{-1}), p_{O2} was seen to decrease gradually along the channels from the inlet to the outlet, showing the uniform power generation on the catalyst layer. At approximately 50 mm along the channel length, p_{O2} began to increase (Fig. 2b), because there was no catalyst layer between 52 and 54 mm [20, 21]. At U_{O2} = 30% (0.055 dm^3 min^{-1}), p_{O2} was also seen to decrease gradually along the channel, with a larger incline than that at 15%. In the three central channels (Channels 2, 3, and 4), p_{O2} decreased in a similar manner. In Channels 1 and 5, p_{O2} was slightly higher than that in the other channels at the same channel length. The difference among the channels can be explained by the geometry of the cathodic reaction area as previously reported [20, 21].

Figure 3 shows the cell resistance under dry nitrogen at 40.5 °C prior to cooling down (a), under dry nitrogen from 40.5 to −10 °C (b), during obtaining plots for the calibration of p_{O2} carried out at −10 °C (c), and before/after the cold startup at −10 °C (d). The data in Fig. 3a–d were obtained in this order. At 40.5 °C, the cell resistance suddenly increased 17 min after the supply of dry nitrogen (Fig. 3a). After 20 min, the cell resistance reached 300 mΩ and the cell cooling started. When the cell was cooled down from 40.5 to −10 °C, the cell resistance increased and converged to 500 mΩ (Fig. 3b). During the acquisition of the calibration curves for the p_{O2} visualization, the cell resistance greatly increased to 2800 mΩ. The increase was large especially at the introduction of gasses with the oxygen concentration of 21–25% (between 154 and 182 min in Fig. 3c). This may be due to the higher efficiency for drugging water from the membrane by penetrated oxygen molecules than by nitrogen molecules, because only the nitrogen purging did not show this high resistance at the same period. When the gas was switched from N_2/N_2 (anode/cathode) to H_2/air, the cell resistance suddenly decreased to 2200 mΩ (Fig. 3d). Then the resistance slowly

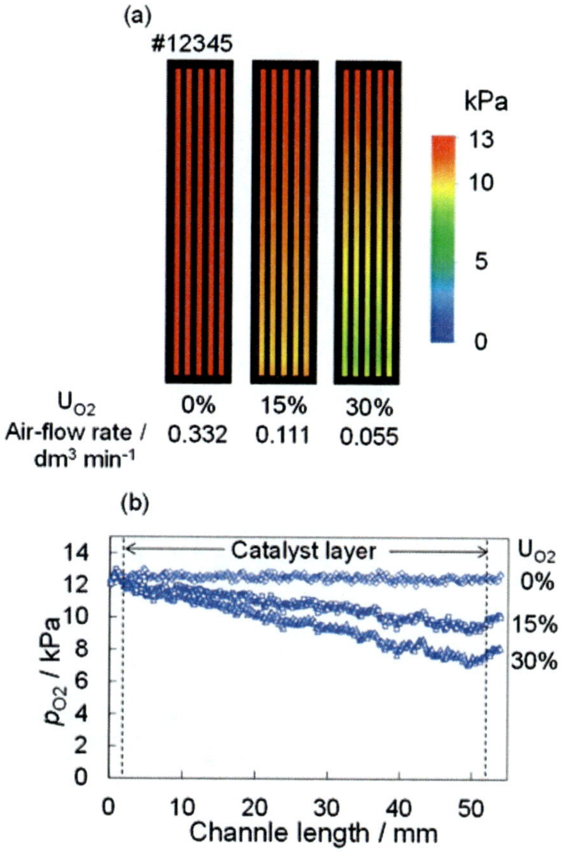

Fig. 2 p_{O2} on the GDL surface visualized in an operating PEFC with channel numbers, #1 to 5 (**a**), and p_{O2} along Channel 3 from the inlet (0 mm) to the outlet (54 mm) (**b**). $U_{O2} = 0$, 15, 30%, and air-flow rate = 0.332, 0.111, and 0.055 $dm^3\ min^{-1}$, respectively. Cell temperature = 80 °C; relative humidity = 80%; H_2 flow rate = 0.200 $dm^3\ min^{-1}$; current density = 0.6 A cm^{-2} except U_{O2} = 0%. The exposure time of the CCD camera was 10 s

decreased because of the formation of water by the cross-linking of gasses across the membrane. On the power generation at 2 mA cm^{-2}, the resistance abruptly increased from 1000 to 2100 mΩ, but it steadily decreased down to 100 mΩ during the power generation.

Time dependences of the cell voltage and the current density in the cold startup at −10 °C are shown in Fig. 4. When the load is initially applied (2 mA cm^{-2}), the voltage was momentarily unstable and reached 0.4 V due to the higher resistance of the MEA. After the cell operation, water and heat were generated at the MEA. The cell voltage therefore slowly increased, because of the water accumulation and the higher temperature within the MEA [1]. Current density was increased to 6, 10, 15, 20, and 30 mA cm^{-2} when the cell voltage reached 0.6 V. The cell voltage began to decrease approximately 1 min after the current was increased to 30 mA cm^{-2}. During the decrease of the cell voltage, the current was increased to 40 and 50 mA cm^{-2}, but the voltage did not recover.

Figure 5 shows p_{O2} on the GDL surface during the cold startup test at −10 °C, where the current density–voltage profile is shown in Fig. 4. At 0 mA cm^{-2} (U_{O2} =

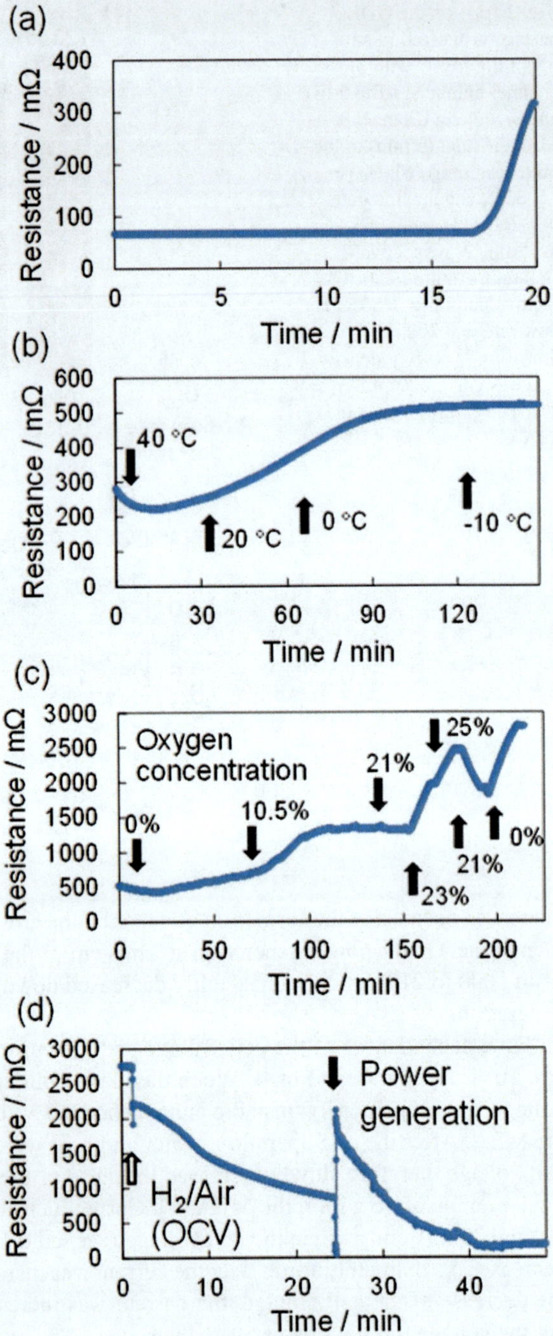

Fig. 3 Cell resistance during the cold startup test. **a** Dry nitrogen to the anode and cathode until the cell resistance became 300 mΩ. Cell temperature = 40.5 °C; nitrogen flow rate = 0.100 $dm^3\ min^{-1}$. **b** Dry nitrogen from 40.5 to − 10 °C at 0.100 $dm^3\ min^{-1}$. **c** Mixtures of oxygen and nitrogen (0–25% oxygen) to the cathode and dry nitrogen to the anode. Cell temperature = −10 °C; mixed gas flow rate = 0.400 $dm^3\ min^{-1}$; nitrogen flow rate of anode = 0.100 $dm^3\ min^{-1}$. **d** During cold startup. Cell temperature = −10 °C; hydrogen-flow rate = 0.100 $dm^3\ min^{-1}$; air-flow rate = 0.030 $dm^3\ min^{-1}$

Fig. 4 Cell voltage and the current density during the cold startup. Cell temperature = −10 °C; dry hydrogen-flow rate = 0.100 dm^3 min^{-1}; dry air-flow rate = 0.030 dm^3 min^{-1}

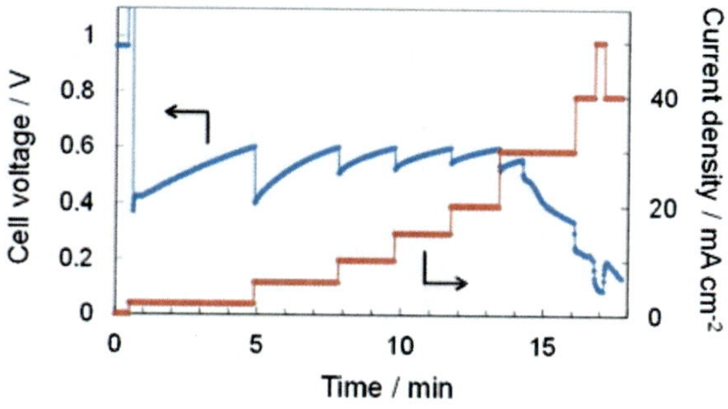

0%, Fig. 5a), before the cold startup, p_{O2} was 21 kPa throughout the channels. During the power generation at 10 mA cm^{-2} (U_{O2} = 2.8%, Fig. 5b), p_{O2} was observed to decrease from the inlet to the outlet. In Channels 1 and 5, p_{O2} decreased smoothly along the channels, but in Channels 2–4, p_{O2} decreased more near the inlet. At the outlet, p_{O2} in all channels became approximately 19.5 kPa. At 20 mA cm^{-2} (U_{O2} = 5.5%, Fig. 5c), p_{O2} decreased smoothly along the channels in Channels 1 and 5. In Channels 2–4, p_{O2} abruptly decreased around 10 mm from the inlet along the channel length. Before and after this location along the channel length, the p_{O2} change was small. Therefore in Channels 2–4, the catalyst layer was most active around 10 mm from the inlet. On the GDL surface near/at the rib walls near the inlets of the Channels 2–4, where the power generation proceeded, some round spots, 0.1–0.3 mm in diameter, are seen in blue and green in Fig. 5c. These irregular spots are ice particles. The ice particles were not confirmed with eyes, so they might not exist on the outmost surface of the GDL, but rather inside the GDL close to the surface. At 30 mA cm^{-2} (U_{O2} = 8.3%, Fig. 5d), p_{O2} decreased around 20 mm along the channels in Channels 1 and 5, and 10 mm in Channels 2–4. The number and the size of ice particles became larger with increasing the current density from 20 to 30 mA cm^{-2} (Fig. 5c, d) in Channels 2–4. Small amounts of ice particles were also seen in Channels 1 and 5. After 90 and 150 s (Fig. 5e, f), p_{O2} changed mainly in the middle of all the channels by the power generation. Neither the number nor the size of ice particles seen at 0 s at 30 mA cm^{-2} (U_{O2} = 8.3%, Fig. 5d) increased at 90 (Fig. 5e) and 150 s (Fig. 5f). When the current density was further increased to 50 mA cm^{-2} (U_{O2} = 13.8%, Fig. 5g), p_{O2} in the central channels (Channels 2–4) changed around 40 mm along the channel length, whereas that in the side channels (Channels 1 and 5) changed at the middle of the channels. Therefore, in the three central channels, the catalyst layer was working only near the outlet at 50 mA cm^{-2}. By the water generation, all the catalyst layers were eventually frozen, and the power generation was stopped (Fig. 5h).

Figure 6a–c show p_{O2} along Channel 3 (red line) and 1 (blue line) from the inlet to the outlet at U_{O2} = 8.3% before the voltage drop (Fig. 5d), at U_{O2} = 8.3% after the voltage drop (Fig. 5f), and at U_{O2} = 13.8% after the voltage drop (Fig. 5g), respectively. The dashed black lines in Fig. 6 represent calculated p_{O2} values assuming a uniform power generation throughout the catalyst layer, a uniform gas distribution

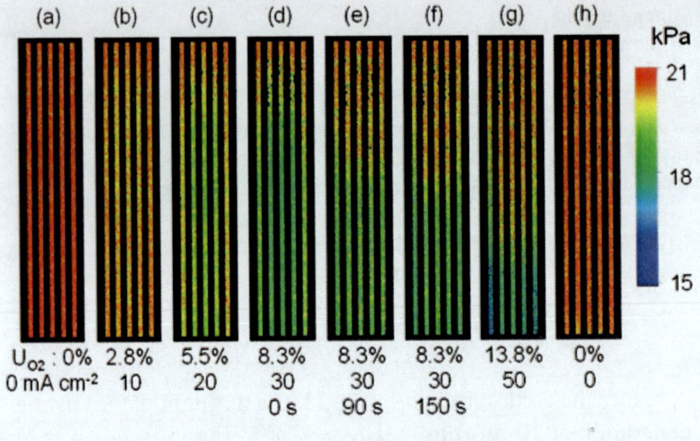

Fig. 5 p_{O2} on the GDL visualized during the cold startup. Current density = 0 (**a**), 10 (**b**), 20 (**c**), 30 mA cm^{-2} at 0 s (before the voltage drop) (**d**), 30 mA cm^{-2} at 90 s (after the voltage drop) (**e**), 30 mA cm^{-2} at 150 s (after the voltage drop) (**f**), 50 mA cm^{-2} (after the voltage drop) (**g**), and 0 mA cm^{-2} (after the test) (**h**); U_{O2} = 0% (**a**), 2.8% (**b**), 5.5% (**c**), 8.3% (**d**)–(**f**), 13.8% (**g**), and 0% (**h**); The exposure times of the CCD camera were 10 s

across the flow channels, the same total pressure throughout the channels, and no water vapor partial pressure at subzero temperature with dry air and hydrogen. In Fig. 6a, p_{O2} in Channel 1 is seen to decrease gradually from the inlet to 30 mm along the channel length, whereas in Channel 3, p_{O2} decreased till 20 mm along the channel length. In both channels, p_{O2} was the same down to the outlet. At the outlet, p_{O2} in Channel 3 (19.0 kPa) was smaller than that in Channel 1 (18.6 kPa), showing the larger oxygen consumption in Channel 3. The estimated value at the outlet is 18.8 kPa, which corresponds to the average value of p_{O2} at the outlet of all the channels. In Fig. 6b, p_{O2} in Channel 1 decreased gradually from the inlet to 40 mm along the channel length, whereas in Channel 3, p_{O2} decreased largely at 30 mm along the channel length. At the outlet, p_{O2} in Channel 3 (19.1 kPa) was now larger than that in Channel 1 (18.6 kPa), therefore, oxygen is consumed more in Channel 1 than in Channel 3. In Fig. 6c, p_{O2} in Channel 1 decreased gradually from the inlet to the outlet, whereas in Channel 3, p_{O2} decreased largely around 40 mm along the channel length. The difference of p_{O2} at the outlet became large at the outlet by 1.4 kPa, showing that Channel 3 became very inactive.

When the load is initially applied, the voltage increase and the resistance decrease proceeded with power generation near the inlet, because of the generated water that enhanced the proton conductivity at the membrane. As the power generation continued, the ice particles were observed near the inlet because of the generated water. Then, the catalyst layer became inactive because of freezing. The active areas continuously moved downwards. At the earlier stages of the power generation, the active areas located in the central three channels, Channels 2–4. This could be explained by the higher temperature in the central part of the cell, keeping the generated heat. The higher power generation in the central three channels caused a

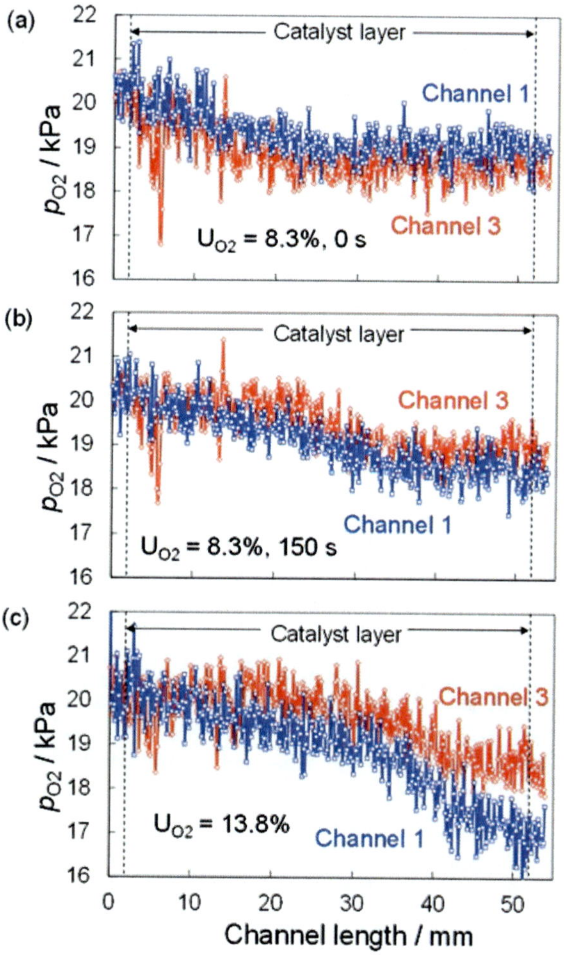

Fig. 6 p_{O2} on the surface of GDL along Channel 3 (red line) and 1 (blue line) from the inlet (0 mm) to the outlet (54 mm) at $U_{O2} = 8.3\%$ before the occurrence of voltage drop (see Fig. 5d) (**a**), at $U_{O2} = 8.3\%$ after the occurrence of voltage drop (see Fig. 5f) (**b**), and at $U_{O2} = 13.8\%$ after the occurrence of voltage drop (see Fig. 5g) (**c**). Calculation data along the channel length are shown as a dashed black line

self-deactivation earlier than the side channels. The side channels were eventually deactivated by freezing, and the power generation stopped.

4 Conclusions

The cold startup test of a PEFC at −10 °C was carried out with visualization of p_{O2} on the GDL surface in five straight channels. During the cooling down of the cell filled with dry nitrogen from 40.5 to −10 °C, the cell resistance increased and converged from 300 to 500 mΩ. The dried mixture gas (N_2/O_2) supplied for the acquisition of the calibration curves for the visualization at −10 °C greatly increased the cell resistance increased to 2800 mΩ. During the cold startup, the current density was increased from 2 to 30 mA cm^{-2} when the cell voltage reached 0.6 V. Before the voltage drop, oxygen was found to be consumed near the inlet. The central three channels generated more power than the side two channels. After the voltage drop, p_{O2} changed mainly in the middle of the channels. Ice particles were also observed near/at the ribs. The side channels became more active than the central channels. When the current density was further increased to 50 mA cm^{-2}, in the three central channels, the catalyst layer was working only near the outlet. Then, all the catalyst layers were frozen, and the power generation was stopped.

Acknowledgements This work was partially supported by SPer-FC and ECCEED'30-FC projects, NEDO, Japan and by Kakenhi, JSPS, Japan.

References

1. A. Pistono, C.A. Rice-York, V. Boovaragavan, J. Electrochem. Soc. **158**(2), B233–B238 (2011)
2. P. Oberholzer, P. Boillat, R. Siegrist, R. Perego, A. Kästner, E. Lehmann, G.G. Scherer, A. Wokaun, J. Electrochem. Soc. **159**(2), B235–B245 (2012)
3. S. Hirakata, T. Mochizuki, M. Uchida, H. Uchida, M. Watanabe, Electrochim. Acta **108**, 304–312 (2013)
4. X.G. Yang, Y. Tabuchi, F. Kagami, C.Y. Wang, J. Electrochem. Soc. **155**, B752–B761 (2008)
5. Q. Guo, Z. Qi, J. Power Sources **160**, 1269–1274 (2006)
6. K. Tajiri, Y. Tabuchi, C.Y. Yang, J. Electrochem. Soc. **154**, B147–B152 (2007)
7. F. Jiang, W. Fang, C.Y. Yang, Electrochim. Acta **53**, 610–621 (2007)
8. E.L. Thompson, J. Jorne, W. Gu, H.A. Gasteiger, J. Electrochem. Soc. **155**, B625–B634 (2008)
9. E.L. Thompson, J. Jorne, W. Gu, H.A. Gasteiger, J. Electrochem. Soc. **155**, B887–B896 (2008)
10. Y. Ishikawa, H. Hamada, M. Uehara, M. Shiozawa, J. Power Sources **179**, 547–552 (2008)
11. Y. Hiramitsu, N. Mitsuzawa, K. Okada, M. Hori, J. Power Sources **195**, 1038–1045 (2010)
12. Z. Miao, H. Yu, W. Wei, L. Hao, Z. Shao, Q. Shen, J. Hou, B. Yi, Int. J. Hydrogen Energy **35**, 5552–5557 (2010)
13. K. Tajiri, Y. Tabuchi, F. Kagami, S. Takahashi, K. Yoshizawa, C.Y. Wang, J. Power sources **165**, 279–286 (2007)
14. J. Ko, H. Ju, Int. J. Hydrogen Energy **39**, 2854–2862 (2014)
15. J.S. Pierre, J. Roberts, K. Colbow, S. Campbell, A. Nelson, J. New Mater. Electrochem. Syst. **8**, 163–176 (2005)
16. S. Hirakata, M. Hara, K. Kakinuma, M. Uchida, D.A. Tryk, H. Uchida, M. Watanabe, Electrochim. Acta **120**, 240–247 (2014)
17. J. Hou, H. Yu, L. Wang, D. Xing, Z. Hou, P. Ming, Z. Shao, B. Yi, J. Power Sources **180**, 232–237 (2008)
18. E.L. Thompson, T.W. Capehart, T.J. Fuller, J. Jorne, J. Electrochem. Soc. **153**(12), A2351–A2362 (2006)
19. J. Li, S. Lee, J. Roberts, Electrochimi. Acta **53**, 5391–5396 (2008)
20. K. Nagase, T. Suga, Y. Nagumo, M. Uchida, J. Inukai, H. Nishide, M. Watanabe, J. Power Sources **273**, 873–877 (2015)
21. K. Nagase, H. Motegi, M. Yoneda, Y. Nagumo, T. Suga, M. Uchida, J. Inukai, H. Nishide, M. Watanabe, ChemElectroChem. https://doi.org/10.1002/celc.201402385.

22. S. Ge, C.Y. Wang, Electrochim. Acta **52**, 4825–4835 (2007)
23. Y. Ishigami, W. Waskitoaji, M. Yoneda, K. Takada, T. Hyakutake, T. Suga, M. Uchida, Y. Nagumo, J. Inukai, H. Nishide, M. Watanabe, J. Power Sources **269**, 556–564 (2014)
24. J. Wang, Energy. Sci. Technol. **2**, 1–12 (2011)

In Silico of Hydrogen Rich Gas from Formic Acid by Steam Reforming Process via DFT Calculation

Mohd Sofi Numin, Khairulazhar Jumbri, and Almila Hassan

Abstract Theoretical molecular behaviour analysis of the hydrogen rich gas production from the steam reforming process of formic acid (HCOOH) has been studied by quantum chemical calculation. The dehydrogenation of formic acid molecules in aqueous and vacuum conditions yields the formation of carbon dioxide (CO_2), hydrogen (H_2) gases, and water (H_2O) molecules. The geometric properties of formic acid, water, carbon dioxide, hydrogen, reactant (HCOOH + H_2O), and product (CO_2 + H_2 + H_2O) molecules were carried out using DFT/B3LYP method in TmoleX software with a def-SV(P) basis set. The method validation was done by comparing the fundamental features of the optimized formic acid geometry with the literatures. The reactivity, stability, electron donating-accepting electron property, and energetic behaviour of every single molecule, reactant, and product were studied and explained by the energy of HOMO and LUMO, band gap energy (ΔE), chemical potential (μ), the global hardness (η), softness (σ), and electronegativity (χ).

Keywords Steam reforming · Formic acid · Dehydrogenation · Quantum chemical calculation

1 Introduction

The vast technological development in this modern era moves together with the high demand for energy in human life because it is a basic human need. Energy and environmental problems are closely related, including air pollution, climate change, water pollution, thermal pollution, and solid water disposal. Burning fossil fuel is the main contributor to the emission of the greenhouse effect. However, among all burning fuels, hydrogen (H_2) is the fuel that is known to be clean when consumed in the fuel cell because the energy is released when hydrogen molecules react with oxygen (O_2) to form water (H_2O) which is harmless to the environment [1]. Steam reforming reaction of hydrocarbon molecules is a promising approach in producing

M. S. Numin · K. Jumbri (✉) · A. Hassan
Department of Fundamental and Applied Sciences, Universiti Teknologi PETRONAS, 32610 Seri Iskandar, Perak, Malaysia
e-mail: khairulazhar.jumbri@utp.edu.my

M. B. Othman et al. (eds.), *Proceedings of the 1st International Conference of New Energy*, Springer Proceedings in Energy, https://doi.org/10.1007/978-981-99-0859-2_5

a hydrogen-rich gas. Formic acid is one of the simplest products formed via the electrochemical reduction of carbon dioxide (CO_2) in the presence of water that can be used as a hydrogen source [2]. The reaction is shown below (Eq. 1):

$$CO_2 + H_2O \rightarrow HCOOH + \frac{1}{2}H_2O \tag{1}$$

The dehydrogenation of formic acid is one of the steam reforming reactions that yield the formation of CO_2 and hydrogen gas [3] (Eq. 2):

$$HCOOH \rightarrow CO_2 + H_2 \tag{2}$$

The high weight percentage of hydrogen (4.3 wt%) in the formic acid is one reason this molecule can be useful for hydrogen storage. Based on molecular weight, at the standard density value of formic acid, 1.22 kg L^{-1} will lead to 53 g of hydrogen per liter of formic acid and 744 g of carbon monoxide per liter of formic acid [4].

A review by Li and Xu [5] on the catalytic activities of metal-nanoparticle-catalyzed hydrogen generation from formic acid shows the exciting research area in the future. The use of highly active metal nanoparticle catalysts can reduce the reaction temperature (25–90 °C) and energy used in the dehydrogenation of formic acid. A study by Yang et al. [6] shows the success and improvement of the selectivity and activity of the surface structure of Pd-alloy catalyst in formic acid dehydrogenation. The addition of a catalyst in the decomposition of formic acid to form hydrogen-rich gas is to reduce the energy use in the reaction. Thus, in-depth geometric energetic behaviour analysis of formic acid decomposition will be a valuable reference for researchers in the future of hydrogen-rich gas production. Nowadays, a quantum chemical calculation is a promising computational chemistry technique for predicting the molecular behaviour of the molecules in terms of energy change and electronic behaviour during the reaction [7, 8]. Therefore, in this work, the density functional theory (DFT) calculation method was used to study the geometric energetic behaviour in the formic acid steam reforming reaction focused on dehydrogenation in the presence of and without water.

2 Method on Computational Chemistry

The molecular structure of each of the molecules in the reactants and products of the steam reforming process reaction were optimized based on two equations below (Eqs. 3 and 4):

$$HCOOH \rightarrow CO_2 + H_2\text{: dehydrogenation without water (Reaction 1)} \tag{3}$$

$$HCOOH + H_2O \rightarrow CO_2 + H_2 + H_2O\text{: dehydrogenation with water (Reaction 2)} \quad (4)$$

The other system that contains only reactant or product molecules for reaction 1 and 2 are also optimized. The DFT calculation is performed by using Tmolex software. The initial coordinates of each molecule were submitted to the Topology Builder (ATB) and Repository Version 3.0 server to obtain the optimized geometry chemical file. The structure of each molecule, reactants, and product were optimized by using hybrid functional B3LYP [9, 10] with def-SV(P).h basis set [11, 12]. The ground-state calculation was inserted under the DFT setting to generate the input file, and the molecule's visualization and parameters derived from the DFT calculation were visualized and calculated by the Tmolex program.

3 Results and Discussions

3.1 *DFT Method Validation*

Method validation is essential in producing accurate results in computational chemistry technique based on the molecular conditions in each ensemble. It is based on comparing the parameters calculated from DFT calculations with the value from the literature (computer simulation or experimentally). The method validation was done by comparing the total energy of each of the molecules after it was optimized. The bond length, $\boldsymbol{r}$ and angle, $\boldsymbol{\delta}$ for the formic acid molecules, were also calculated and compared with the literature. Figure 1 shows the optimized structure of formic acid with the bond length value. The calculated bond length and angle comparison with the values from the literature were shown in Table 1 and indicated excellent agreement with a low relative error (0.003–2.24%). The energy for each molecule was also calculated and compared with the value from the literature for a method validation purpose (Table 2). The low relative error (0.12–0.94%) for the energy value calculated also shows a good agreement with the literature, indicating the method used was validated and can be used for further analysis.

3.2 *Quantum Chemical Parameters*

Geometry Optimization. Geometry optimization is a process of lowering the energy to find an atomic arrangement of molecules which makes it most stable [16]. The primary purpose of the quantum chemical calculation is to calculate molecules' structures, interactions, and properties in the most stable state [17]. The minimum energy for each molecule, reactant and product is calculated after optimization. The bond length and angle for each molecule were also calculated. The parameters are all

Fig. 1 Optimized structure of formic acid

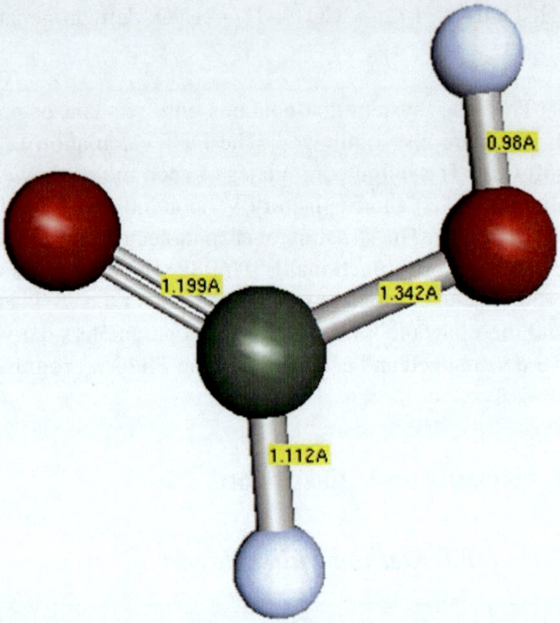

Table 1 Comparison of bond length (Å)) and angle (°) of optimized structure of acetic acid

Bond length, *r* (Å) and angle, *δ* (°)	Calculated	Experimental [13]	Relative error (%)	Computer simulation [14]	Relative error (%)
r (C–H)	1.112	1.106	0.54	1.107	0.45
r (C–O)	1.199	1.217	1.48	1.213	1.15
r (C–OH)	1.342	1.361	1.40	1.356	1.03
r (H–C–OH)	0.980	0.984	0.41	0.983	0.31
δ (H–C–OH)	107.719	109.1	1.27	109.3	1.45
δ (C–O–H)	109.296	107.3	1.86	106.9	2.24
δ (O–C–OH)	125.284	123.4	1.53	125.3	0.01

Table 2 Comparison of total energy of optimized structure of formic acid with other literature

Molecules	Total energy value (a.u)		
	Calculated	Literature	Relative error (%)
Formic acid	−189.519	−188.762 [15]	0.40
Water	−76.302	−76.409 [7]	0.14
Hydrogen gas	−1.164	−1.175 [7]	0.94
Carbon dioxide	−188.361	−188.581 [7]	0.12

Table 3 Total energy, bond length, and bon angle for each molecule, reactants, and products after DFT/B3LYP optimization with def-SV(P) basis set

Parameters	Total energy (a.u)	Bond length, r (Å)	Bond angle, d (°)
Formic acid	−189.519	C–H = 1.112, C–O = 1.199, C–OH = 1.342, O–H = 0.980	H–C–OH = 107.719, C–O–H = 109.296, O–C–OH = 125.284
Water	−76.302	O–H = 0.972	H–O–H = 104.363
Hydrogen gas	−1.164	H–H = 0.765	–
Carbon dioxide	−188.361	C–O = 1.163	O–C–O = 180
Formic acid + water (Reactant)	−265.848	–	–
Carbon dioxide + hydrogen gas (Product)	−189.525	–	–
Carbon dioxide + hydrogen gas + water (Product)	−265.837	–	–

tabulated in Table 3. After optimization, the total energy of formic acid is −185.519 a.u, which means it is the lowest energy where the molecule reaches its most stable state. Besides, for H_2O, CO_2, and H_2, the total energy value is −76.302, −188.361, and −1.164 a.u, respectively.

The total energy before and after reaction for both reactions (with and without water) is also calculated, and it is found that the total energy without water (Reaction 1) for reactant is −185.519 and after reactions are −189.525 a.u. The product side of this reaction shows a −8.000 a.u interaction energy value, indicating the energy reduced when both molecules of products (H_2 and CO_2) interact with each other. On the other hand, for reaction with the presence of water (Reaction 2), the total energy before and after reactions are −265.848 and −265.837 a.u, respectively. DFT calculation also calculated the interaction energy for both reactant and product. For the reactant, the −0.027 a.u of interaction energy value shows the energy reduced when HCOOH and H_2O react at −0.027 a.u, while the energy reduced for the product molecules ($CO_2 + H_2 + H_2O$) to interact is 0.011 a.u.

HOMO–LUMO and Bandgap Energy. The reactivity of each molecule in the steam reforming process with and without the presence of water was investigated by calculating the highest occupied molecular orbital (E_{HOMO}) energy, lowest unoccupied molecular orbital (E_{LUMO}) energy, and bandgap energy ($\Delta E = E_{LUMO} - E_{HOMO}$). Figure 2 shows the geometric representation of HOMO–LUMO for each optimized molecule, reactant, and product (Reactions 1 and 2). The HOMO region represents a part where the molecules can donate their electron/s, while LUMO represents a region that can accept electron/s [18, 19].

Figure 3 shows the HOMO, LUMO and bandgap energies for each molecule, reactant, and product. The energy of HOMO and LUMO for formic acid are −0.2926 and 0.0007 a.u, respectively. Among all molecules, water has the highest HOMO energy with a value of −0.2884 a.u. This indicates the ability of water

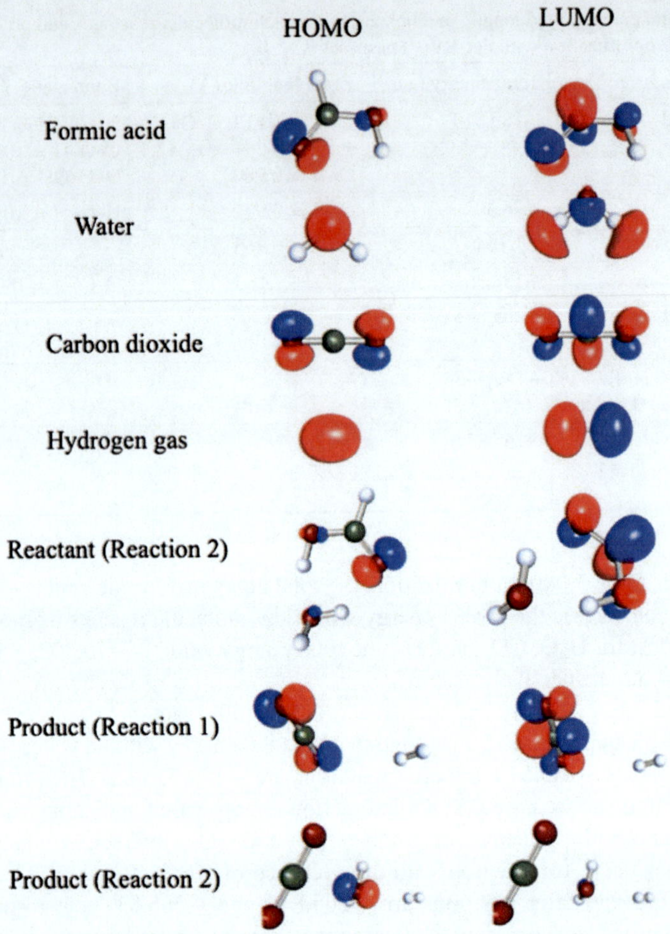

Fig. 2 HOMO and LUMO for each molecule, reactant, and product after DFT/B3LYP optimization with def/SV(P) basis set

molecules to act as a nucleophile to donate an electron to the other reactant for the reaction to occur with a low ionization potential [7]. On the other hand, the LUMO energy of water molecules is low (0.0421 a.u), indicating the capacity of high electron affinity and lower electron acceptance ability. The high electron acceptance ability of water molecules is compatible with the high HOMO value of formic acid with a high electron donating ability resulting in the high reactivity of the steam reforming process in the presence of water. The work from [3] supported the results where adding water to the formic acid decomposition process reduced the reaction activation energy and increased its reactivity. Both reactant molecules (formic acid and water) also have a low bandgap energy value, representing the most reactive molecules compared to carbon dioxide and hydrogen gas. The high bandgap energy value of

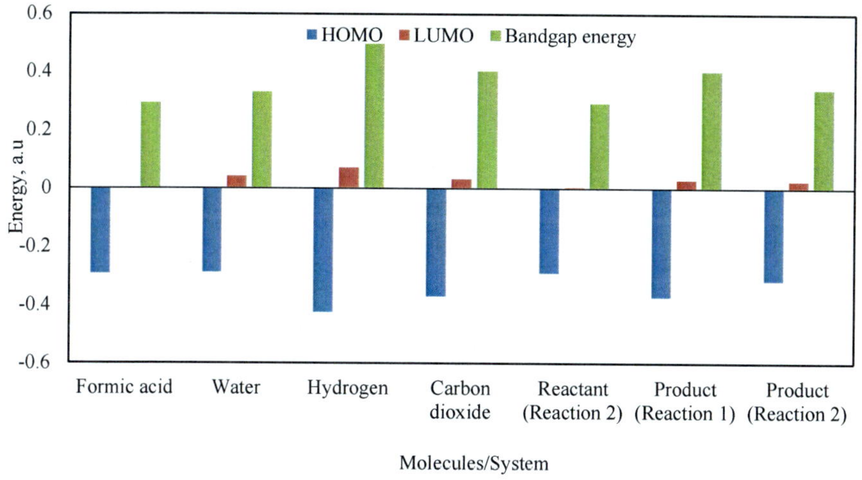

Fig. 3 HOMO, LUMO, and bandgap energies for each molecule, reactant, and product after DFT/B3LYP optimization with def-SV(P) basis set

carbon dioxide and hydrogen gas does not affect the reaction reactivity because the product is formed after the reaction. The energetic behaviours (HOMO, LUMO, and bandgap energy) of the reactant system and products system are also calculated and show that the product side has the highest stability compared to reactants for both reactions, indicating the high reactivity of the reactant to form a product from high energy to a low energy state.

Global Hardness, Global Softness, Chemical Potential, and Electronegativity. The reactivity of the molecules was also investigated with other quantum parameters such as global hardness (η), global softness (σ), chemical potential (μ), and electronegativity (χ). Global hardness and softness are related to the HOMO and LUMO energies of the molecules, which indicates the chemical stability of the molecules [20]. According to Hizaddin et al. [21], global hardness is related to the HOMO and LUMO energy gap of the molecules/systems with an expression (Eq. 5):

$$\eta = \frac{E_{\text{LUMO}} - E_{\text{HOMO}}}{2} \tag{5}$$

On the other hand, global softness is the reciprocal of a global hardness, which can describe the capability of an atom or group to receive electrons, was calculated by using the following equation [22] (Eq. 6):

$$\sigma = \frac{1}{\eta} \tag{6}$$

Chemical potential is the first derivative of the overall energy regarding the number of electrons in a molecule, where electrons will flow from the high chemical potential to the low chemical potential [23, 24]. Electronegativity explains the capability of any molecule to attract electrons [25]. The expression of chemical potential and electronegativity are as follows (Eqs. 7 and 8):

$$\mu = -\frac{I + A}{2} \tag{7}$$

$$\chi = \frac{I + A}{2} \tag{8}$$

The calculated global hardness, global softness, chemical potential, and electronegativity value are tabulated in Table 4. Formic acid has the lowest hardness value (0.1467 a.u) and highest softness value (6.8190 a.u). These low and high hardness and softness values indicate that this molecule is easily polarizable due to the small resistance to alter its electronic arrangement and ease in transferring the charge to this molecule [7]. Water molecule also has the same property as formic acid with a lower softness value (0.1653 a.u) and high softness value (6.0514 a.u) compared to hydrogen gas and carbon dioxide. DFT method was also used to calculate the hardness and softness value for the reactant and product sides, and it was found that the reactant side for both reactions 1 and 2 has a low hardness and higher softness compared to the product side. This is important since the low hardness value and high softness value of reactant indicate the reactivity of the reactant molecules to undergo a rapid steam reforming process.

Electronegativity indicates the ability of any molecule to attract electrons [26]. Among all molecules, the water molecule has the lowest electronegativity value (0.1232 a.u), which indicates this molecule's weak electron acceptance properties and has a better electron donor property. It is followed by formic acid, carbon dioxide, and hydrogen gas. The higher electronegativity value for formic acid compared to

Table 4 Global hardness, global softness, chemical potential, and electronegativity of each molecule, reactant, and product after DFT/B3LYP optimization with def-SV(P) basis set

Molecules	Global hardness, η (a.u)	Global softness, σ (a.u)	Chemical potential, μ (a.u)	Electronegativity, χ (a.u)
Formic acid	0.1467	6.8190	−0.1460	0.1460
Water	0.1653	6.0514	−0.1232	0.1232
Hydrogen gas	0.2481	4.0306	−0.1761	0.1761
Carbon dioxide	0.2019	4.9542	−0.1683	0.1683
Reactant (Reaction 2)	0.1469	6.8097	−0.1427	0.1427
Product (Reaction 1)	0.2014	4.9652	−0.1700	0.1700
Product (Reaction 2)	0.1704	5.8685	−0.1455	0.1455

water molecule represent that the oxygen atom (HOMO region) in the water molecule will act as an electron donor to the empty orbital of the formic acid molecule due to its high electron acceptance ability compared to water. Reactions 1 and 2 for the steam reforming process show that the electronegativity of the reactant side is lower than the product side. This indicates that the reactant side is a good electron donation system compared to the product for both reactions 1 and 2. The chemical potential for each molecule, reactant, and product is also calculated. It is a negative value of electronegativity that explains the ability of an electron to escape from the molecules. Water molecules have the highest chemical potential value compared to the other molecules, followed by formic acid, carbon dioxide, and hydrogen gas. Water and formic acid molecule, which are also the reactant molecules in the steam reforming process, have a greater tendency to make the electron escape and donate the electron compared to the product side of the reaction based on chemical potential.

4 Conclusion

The geometric optimization of dehydrogenation of formic acid in the steam reforming process is investigated by the DFT/B3LYP method with a def-SV(P) basis set using Tmolex software. The method was successfully validated by comparing the total energy of 4 molecules involved in the reaction without and with the presence of water with the value from the literature. The bond length and angle of the formic acid were also calculated and compared, and the low relative error proves the method's validity. The high value of $\boldsymbol{E}_{\mathbf{HOMO}}$ for water molecules that give an excellent electron donating tendency and high $\boldsymbol{E}_{\mathbf{LUMP}}$ of formic acid that has the electron acceptance property shows the reactivity of the steam reforming process is higher with the presence of water molecules. The global hardness, global softness, chemical potential, and electronegativity value also show the greater tendency of water molecules to donate electrons to the LUMO region of formic acid in the formic acid decomposition process with the presence of water. The in-depth study on the steam reforming process assumed that the dehydrogenation process of formic acid to form hydrogen gas is more favourable in the presence of water molecules. These results also show an excellent agreement from the literature where water molecules increase the reaction rate of formic acid decomposition. The further in-depth study via computational chemistry for the other organic acid to form hydrogen-rich gas molecules, especially with the presence of a metal catalyst, would be a valuable reference for hydrogen gas production in the future. The computational chemistry method will be cost and time-friendly and can reduce the amount of chemicals and energy used compared to investigating experimentally.

Acknowledgements This research was funded by the PETRONAS Research Sdn Bhd and Universiti Teknologi PETRONAS (GR&T UTP) Collaboration (grant number 015MD0-085) and Yayasan Universiti Teknologi PETRONAS-Fundamental Research Grant (YUTP-FRG) grant number 015LC0-409.

References

1. Hydrogen Fuel Energy, Office of Energy Efficiency & Renewable Energy, https://www.energy.gov/eere/fuelcells/hydrogen-fuel-basics. Last accessed 8 Sept 2022
2. A. Rahbari, M. Ramdin, L.J.P.V.D. Broeke, T.J.H. Vlugt, Combined steam reforming of methane and formic acid to produce syngas with an adjustable H_2:CO ratio. Ind. Eng. Chem. Res. **57**, 10663–10674 (2018)
3. N. Akiya, P.E. Savage, Role of water in formic acid decomposition. AIChE J. **44**(2), 405–415 (1998)
4. K. Müller, K. Brooks, T. Autrey, Hydrogen storage in formic acid: a comparison of process options. Energy Fuels **31**, 12603–12611 (2017)
5. Z. Li, Q. Xu, Metal-nanoparticle-catalyzed hydrogen generation from formic acid. Acc. Chem. Res. **50**, 1449–1458 (2017)
6. Y. Yang, H. Xu, D. Cao, X.C. Zeng, D. Cheng, Hydrogen production via efficient formic acid decomposition: engineering the surface structure of pd-based alloy catalysts by design. ACS Catal. **9**(1), 781–790 (2019)
7. M.A. Hossain, J. Jewaratnam, A. Ramalingam, J.N. Sahu, P. Ganesan, A DFT method analysis for formation of hydrogen rich gas from acetic acid by steam reforming process. Fuel **212**, 49–60 (2018)
8. V. Jaiswal, R.B. Rastogi, J.L. Maurya, P. Singha, A.K. Tewari, Quantum chemical calculation studies for interactions of antiwear lubricant additives with metal surfaces. RSC Adv. **26**, 13438–13445 (2014)
9. A.D. Becke, Density-functional theory. III. The role of exact exchange. J. Chem. Phys. **98**(7), 5648–5652 (1993)
10. A.D. Becke, Density functional calculations of molecular bond energies. J. Chem. Phys. **84**(8), 4524–4529 (1986)
11. M.D. Hanwell, D.E. Curtis, D.C. Lonie, T.V.E. Zurek, G.R. Hutchison, Avogadro: an advanced semantic chemical editor, visualization, and analysis platform. J. Chemicformatics **4**, 17 (2012)
12. C. Steffen, K. Thomas, U. Huniar, A. Hellweg, O. Rubner, A. Schroer, Software news and updates TmoleX-a graphical user interface for TURBOMOLE. J. Comput. Chem. **31**, 2967–2970 (2010)
13. A. Almenningen, O. Bastiansen, T. Motzfeldt, A reinvestigation of the structure of monomer and dimer formic acid by gas electron diffraction technique. Acta Chemica Scaninavica **23**, 2848–2864 (1969)
14. S. Duan, Y.F. Ji, P.P. Fang, Y.X. Chen, X. Xu, Y. Luo, Z.Q. Tian, Density functional theory study on the adsorption and decomposition of the formic acid catalyzed by highly active mushroom-like Au@Pd@Pt tri-metallic nanoparticles. Phys. Chem. Chem. Phys. **15**, 4625–4633 (2013)
15. F.I.D.C. Moule, D.J. Clouthier, Laser spectroscopic and quantum chemical studies of the lowest excited states of formic acid. J. Phys. Chem. **94**(6) (1990)
16. R. Stephane, *Molecular Geometry Optimization: Algorithms, Encyclopedia of Applied and Computational Mathematics* (Berlin, Heidelberg, Springer, 2015), pp. 957–961
17. E. Shakerzadeh, Theoretical investigations of interactions between boron nitride nanotubes and drugs, in *Boron Nitride Nanotubes in Nanomedicine* (Boston, William Andrew Publishing, 2016), pp. 59–77
18. J. Fang, J. Li, Quantum chemistry study on the relationship between molecular structure and corrosion inhibition efficiency of amides. J. Mol. Struct. (Thoechem) **593**(1–3), 179–185 (2002)
19. G. Bereket, E. Hur, C. Ogretir, Quantum chemical studies on some imidazole derivatives as corrosion inhibitors for iron in acidic medium. J. Mol. Struct. (Thoechem) **578**(1–3), 79–88 (2002)
20. B. Kosar, C. Albayrak, Spectroscopic investigations and quantum chemical computational study of (E)-4-methoxy-2-[(p-tolylimino)methyl]phenol. Spectrochim. Acta Part A Mol. Biomol. Spectrosc. **78**(1), 160–167 (2011)
21. H.F. Hizaddin, R. Anantharaj, M.A. Hashim, A quantum chemical study on the molecular interaction between pyrrole and ionic liquids. J. Mol. Liq. **94**, 20–29 (2014)

22. J.H. Henriquez-Roman, L. Padilla-Campos, M.A. Paez, J.H. Zagal, M.A. Rubio, C.M. Rangel, J. Costamagna, G. Cardenas-Jiron, The influence of aniline and its derivatives on the corrosion behaviour of copper in acid solution: a theoretical approach. J. Mol. Struct. (Thoechem) **757**(1–3), 1–7 (2005)
23. C.G. Zhan, A.J. Nichols, D.A. Dixon, Ionization potential, electron affinity, electronegativity, hardness, and electron excitation energy: molecular properties from density functional theory orbital energies. J. Phys. Chem. A **107**, 4184–4195 (2003)
24. R.G. Pearson, Absolute electronegativity and hardness correlated with molecular orbital theory. Proc. Natl. Acad. Sci. **83**(22), 8440–8441 (1986)
25. R.G. Pearson, Absolute electronegativity and hardness: application to inorganic chemistry. Inorg. Chem. **27**(4), 734–740 (1988)
26. J. He, L. Hu, Y. Tang, H. Li, P. Yang, Z. Li, Adsorption features and photocatalytic oxidation performance of M1/3NbMoO6 (M = Fe, Ce) for ethyl mercaptan. R. Soc. Chem. **4**(43), 22334–22341 (2014)

The Effect of Second Metals Towards Physicochemical Properties of Nickel-Based Catalyst Supported on Reduced Graphene Oxide for Hydrogenation of Carbon Dioxide into Methane

Nur Diyan Mohd Ridzuan, Maizatul Shima Shaharun, Mahaletchimi Murugan, Nur Natasha Bintang Mohd Jad, and Siti Nur Azella Zaine

Abstract The continuous carbon dioxide, CO_2 emission into the atmosphere has caused climate change and other environmental issues. As a part of the effort towards a sustainable future, hydrogenation of CO_2 into methane is seen as a useful method to recycle CO_2. This Sabatier reaction can be conducted by reacting hydrogen gas, H_2 with CO_2 under the presence of a Ni-based catalyst which can be enhanced under the presence of support and a second metal. In this work, a nickel-based catalyst supported by reduced graphene oxide nanosheets was synthesized and the influence of the addition of the second metal was studied. Ni/rGO, Ni-Co/rGO, and Ni-Cu/rGO were synthesized using the incipient wetness impregnation method and characterized using X-ray diffraction (XRD), Surface Area and Porosity Analysis (SAP) and H_2 temperature-programmed reduction (H2-TPR). The purpose of adding second metals is to determine the effect on their physicochemical properties such as crystallinity, surface area, dispersion, and reducibility of catalyst for hydrogenation of CO_2. Fixed loading of the metal was used with 15 wt.% nickel (Ni) catalyst and 5 wt.% second metals (M = Co and Cu). Based on the correlation result studies, adding the second metal shows good physicochemical properties which enhance reducibility and reduce surface area compared to other synthesized catalysts which indicates that it has a strong catalytic activity in methanation. Ni-Cu/rGO recorded the lowest reduction temperature which reflects the highest reducibility due to the smaller crystallite size,

N. D. Mohd Ridzuan · M. S. Shaharun (✉) · M. Murugan
Department of Fundamental and Applied Sciences, Universiti Teknologi PETRONAS, 32610 Perak, Seri Iskandar, Malaysia
e-mail: maizats@utp.edu.my

N. D. Mohd Ridzuan
e-mail: nurdiyan.ridzuan@utp.edu.my

N. N. B. Mohd Jad · S. N. A. Zaine
Department of Chemical Engineering, Universiti Teknologi PETRONAS, 32610 Perak, Seri Iskandar, Malaysia

M. B. Othman et al. (eds.), *Proceedings of the 1st International Conference of New Energy*, Springer Proceedings in Energy, https://doi.org/10.1007/978-981-99-0859-2_6

lower surface area, and higher dispersion which was suggested by XRD and SAP results.

Keywords Reduced graphene oxide · Bimetallic catalyst · CO_2 hydrogenation

1 Introduction

The greenhouse gases (GHGs) which are responsible for the warming of the earth comprise approximately 82% carbon dioxide, CO_2. Apparently, the natural equilibrium of CO_2 is being disturbed as the concentration of anthropogenic sources increases over the years. It primarily stemmed from the burning of fossil fuel and coal as well as the industrial sector [1]. In order to reduce the release of CO_2 into the atmosphere, studies have been focusing on two approaches; to capture and store CO_2 or to recycle the CO_2 into valuable energy-bearing compounds [2]. The latter is more assuring because it can reduce CO_2 emissions which is more profitable for the industries involved with environmental issues [3]. Hydrogenation of CO_2 into value-added products such as methanol, methane, and ethanol can be seen as a promising solution to reduce CO_2 emissions. CO_2 methanation, also known as the Sabatier reaction (Eq. 1), is a reaction in which CO_2 is reacted with hydrogen gas, H_2 and CH_4 is produced as the energy carrier carried under the presence of catalysts.

$$CO_2(g) + 4H_2(g) \rightarrow CH_4(g) + 2H_2O(g) \quad (1)$$

In previous studies, metal catalysts from group VIIB such as palladium, ruthenium, and nickel are commonly reported as active catalysts for this reaction. Among these metals, nickel, Ni are widely used due to their high performance over cost ratio [4]. The characteristics and performances of the catalysts are influenced by several variables such as calcination temperature [5], synthesis method [6], metal loading [7], and supporting materials [8, 9]. Even though Ni nanoparticles are active for carbon oxide methanation, most of the studies use Ni nanoparticles supported on high surface area materials [10]. This is because, under high-temperature condition during CO_2 methanation, Ni catalyst may be deactivated due to the sintering of Ni particles, formation of mobile nickel sub-carbonyls, and formation of carbon deposits [9, 11]. Upon supporting the metal on support, the sintering of nanoparticles can be reduced by establishing a physical anchor between the support and Ni particles [12]. Besides, due to the fact that the adsorption of CO_2 onto the catalyst is a critical step in the reaction, the addition of support can improve the adsorption of CO_2 [2]. Finally, the presence of support can improve the dispersion of Ni nanoparticles and in turn, will increase the active sites for the reaction to occur. Apart from incorporating metal on the support, the properties and performance of nickel catalyst could be enhanced synergistically by adding promoters or second metal to the catalyst. Different types of second metals such as Fe [13], Zr [2], Pd [14], and Mg [11, 15] have been incorporated in the Ni catalyst system and were found to be involved in the. To improve the efficiency of CO_2 methanation, second metals including magnesium, iron, copper, and cobalt are used together with Ni as a bimetallic catalyst.

Herein, in this study, a bimetallic catalysts system supported on reduced graphene oxide (rGO) is synthesized and characterized to study its physicochemical properties and potential impact towards CO_2 methanation. rGO is used as catalyst support due to its high surface area and porosity. The presence of an oxygen group on the support will cause the support to have higher acidity which may result in acid–base interaction with nickel catalysts; hence reducing metal sintering and increasing the dispersion and reducibility of nickel metal. The second metals used are cobalt, Co, and copper, Cu. The aim of the addition of second metal is to improve the activity for reverse water gas shift reaction at moderate temperatures [16]. Previous studies show that the addition of second metal can also potentially minimize the CO byproduct by inhibiting CO desorption as an intermediate. Hence, the selectivity towards the product will increase. This can be done by studying the physicochemical properties such as nickel dispersion, particle size, and reducibility of the catalyst. It is expected that the incorporation of second metal will improve the properties of the catalysts and potentially enhance their catalytic activity.

2 Methodology

2.1 Materials

Graphite (20 μm), sulphuric acid, H_2SO_4, (95–97%), potassium permanganate, $KMnO_4$, and nickel nitrate hexahydrate, $NiNO_3{\cdot}6H_2O$, copper (II) nitrate trihydrate, $Cu(NO_3)_2{\cdot}3H_2O$, cobalt (II) nitrate hexahydrate, $Co(NO_3)_2{\cdot}6H_2O$, hydrogen peroxide, H_2O_2 (30%), hydrazine hydrate, N_2H_4 (80%), and sodium nitrate, $NaNO_3$ were purchased from Merck. Hydrochloric acid, HCl (37%), was obtained from Baker. All the solutions were prepared using distilled water.

2.2 Synthesis of Catalysts

Ni/rGO and Ni-M/rGO (M=Cu, Co) catalysts were synthesized via the incipient wetness impregnation method. Initially, metal precursors (15 wt% Ni and 5 wt% M to rGO) were added dropwise to rGO support using an aqueous solution prepared by dissolving the required amount of $Ni(NO_3)_2{\cdot}6H_2O$. The volume of impregnating solution was adjusted to the volume of rGO. The impregnated sample was left for 20 h. Then, it was dried at 80 °C for 12 h in the oven before calcined at the optimized temperature of 400 °C for 2 h. The catalysts were labelled as Ni/rGO and Ni-M/rGO where M is the second metal.

2.3 X-Ray Diffraction Analysis

XRD analysis was carried out to examine the crystallinity of samples. X-ray diffractograms for the catalysts were recorded on a powder diffractometer (X'Pert^3 Powder & Empyrean, PANalytical) with Cu Kα radiation source between 2θ of 5° to 90° (scanning step of 0.01°/step) on continuous scanning. The quantitative characteristics of the catalysts were calculated using formulas (2) and (3). Quantitative analysis of the X-ray diffractogram was performed using OriginLab 8.5 software where fitting was made using the Gaussian mathematical function.

$$\lambda = 2d\ \sin\theta \tag{2}$$

$$D = K_{(111)}\lambda/B\ \cos\theta \tag{3}$$

where λ is the wavelength of Cu Kα radiation (1.5418 Å), d is the interplanar distance, D is the NiO crystallite size, and $K_{(111)}$ is the Scherrer constant for (111) peak of NiO (0.855) and B is full width at half maxima of peaks.

2.4 Surface Area and Porosity Analysis

Surface area and porosity (SAP) analysis of all catalysts was performed using Micromeritics ASAP 2020 analyser with N_2 as adsorbate. The surface area was determined from BET model, pore size and volume were determined from BJH method and micropore analysis was carried out using t-plot analysis.

2.5 H_2-TPR Analysis

Hydrogen temperature-programmed reduction (H_2-TPR) of the catalysts was carried out using TPD/R/O 110 MS equipped with a thermal conductivity detector (TCD) to evaluate the reducibility and metal-support interaction of the catalysts. The pretreatment was conducted from 20 to 150 °C with a heating rate of 20 °C·min^{-1} and a holding time of 60 min in N_2 flow (20 mL·min^{-1}). The analysis was performed in temperature range of 50–800 °C at heating rate of 10 °C·min^{-1} and holding time of 30 min in 5 vol.% H_2/N_2 ratio gas flow (20 mL·min^{-1}).

3 Results and Discussion

3.1 X-Ray Diffractogram of Catalysts

X-Ray diffractogram of catalysts was obtained and shown in Fig. 1 below. As discussed in a previous report [17], the prominent peak at ~26° is assigned to (002) peak of rGO. With the impregnation of 15% Ni onto rGO, reflections for the Ni phase were consistent with those for NiO at 37°, 43°, 62°, 75°, and 79° which correspond to (111), (200), (220), (311) and (222) diffraction planes [18]. The presence of these peaks indicates the success of the wetness impregnation method to incorporate active metal onto rGO support. The absence of a secondary phase or impurities peak is noted from the diffractogram, indicating sample purity.

When 5% of cobalt is added, no discernible diffraction peak for this metal is observed. This might be due to the amorphous state or tiny particles of cobalt produced from the synthesis. This too could be attributed to the superposition of metal oxide CoO with the NiO peaks [15, 19]. In contrast, upon the addition of Cu, a shoulder peak at ~35° was observed in Ni–Fe/rGO and Ni-Cu/rGO. In terms of the quantitative aspect, the addition of a second metal to a nickel-based catalyst affects the size of nickel crystallites as well as its interlayer spacing. As can be seen from Fig. 1, the slight right shift of the (200) Ni peak around was observed when copper is added. This is related to the higher Bragg's angle; hence, lower interplanar spacing of the catalyst, based on the relation in Eq. 2. The size of the nickel crystallites shrank in Ni-Cu/rGO, resulting in greater nickel dispersion. This is proven based on the broader width of the (200) peak based on the relation in Eq. 3. On the other hand, the addition of cobalt resulted in bigger interplanar spacing due to the slight left shift and smaller width of the (200) peak.

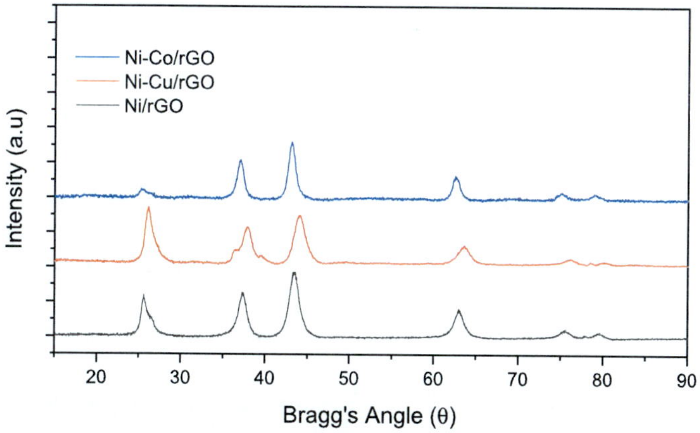

Fig. 1 X-Ray diffractogram of Ni/rGO and Ni-M/rGO catalysts

3.2 Surface Area and Porosity

From Fig. 2, it can be seen that the isotherm of all catalysts exhibits Type IV and possess an H3 hysteresis loop according to IUPAC classification [20]. This indicates that the sample has pores dominated in the mesoporous range (2–50 nm) while the H3-type hysteresis loop represents aggregates of plate-like particles giving rise to slit-shaped pores [21]. The hysteresis loop occurs due to the capillary condensation in the mesopores. The surface area and porosity of the catalysts are tabulated in Table 1. The presence of pores resulted in an increase in surface area and can serve as an active site for the reaction. Based on Table 1, the surface area of the catalysts follows the order of Ni-Co/rGO < Ni/rGO < Ni-Cu/rGO. Upon the addition of Cu as the promoter, the surface area, and pore volume increase meanwhile the addition of Co causes a reduction in surface area and pore volume. This supports the observation of the X-ray diffractogram that suggests cobalt and nickel solid solution on the rGO surface which has a bigger size compared to NiO. As a result, the surface area becomes higher due to bigger particle size. On the other hand, Ni-Cu/rGO shows better textural properties probably due to its low crystallite size which resulted in lower particle size and better dispersion. With that, this explains the higher surface area and porosity of the catalyst with a smaller pore diameter.

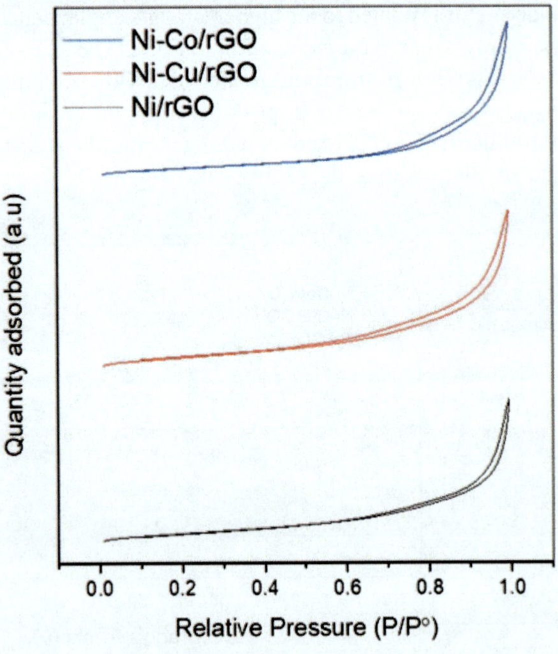

Fig. 2 N_2-sorption isotherm of catalysts

Table 1 Textural properties of catalysts

Catalyst	S_{BET} ($m^2\ g^{-1}$)	Pore volume ($cm^3\ g^{-1}$)	Pore diameter (nm)
Ni/rGO	111.13	0.44	13.64
Ni-Cu/rGO	118.98	0.42	12.59
Ni-Co/rGO	95.51	0.46	16.54

3.3 Reduction Behaviour and Metal-Support Interaction

Temperature programmed reduction (TPR) analysis using H_2 was performed to study the relation between reducibility and metal-support interaction. TPR profiles of the catalysts are presented in Fig. 3 whereas the T_{red} and H_2 consumption is tabulated in Table 2. The profile provided the amount of H_2 gas consumed by the catalysts at different temperatures caused by the redox reaction between H_2 and NiO with the following equation:

$$NiO_{(s)} + H_{2(g)} \rightarrow Ni^{\circ}_{(s)} + H_2O_{(g)} \quad [22]$$

From Fig. 3, the Ni/rGO catalyst shows a strong signal at a temperature of 378 °C caused by the reduction of Ni^{2+} species that are interacting with rGO support [23, 24]. The higher the reduction temperature, the higher the amount of energy needed to overcome the interaction between NiO and rGO [25]. Even though the interaction is important to avoid Ni leaching or agglomeration during the reaction, too strong interaction can be less favourable as it will be harder to be reduced for

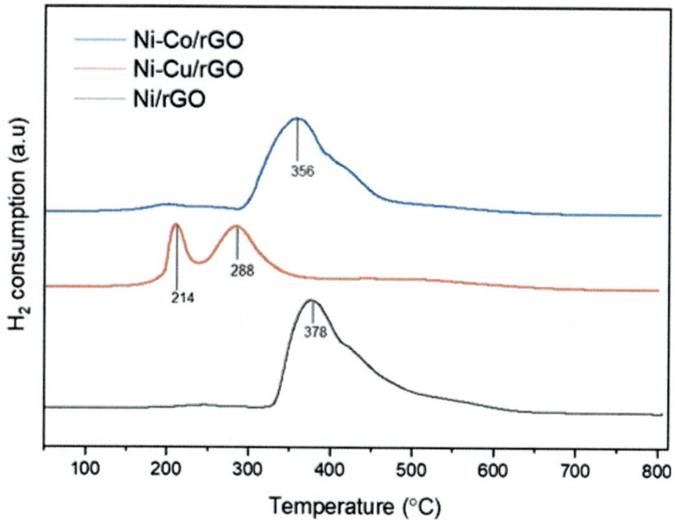

Fig. 3 H_2-TPR of Ni/rGO and Ni-M/rGO catalysts

Table 2 Reduction temperature and H_2 consumption of catalysts

Catalyst	T_{red} (°C)	H_2 Consumption (mmol g_{cat}^{-1})
Ni/rGO	378	5.84
Ni-Cu/rGO	214, 288	6.76
Ni-Co/rGO	356	9.00

catalyst activation. In addition, it is worth noting that the reduction temperature of Ni/rGO catalysts is relatively lower than other catalysts including Ni/Al_2O_3 [26–28], Ni/SiO_2 [22] and Ni/ZrO_2 [29] which reflects the higher reducibility of Ni^{2+} in Ni/rGO catalysts.

Interestingly, upon the addition of the second metal, the reduction temperature further decreases which shows higher reducibility and weaker metal-support interaction.

Ni-Cu/rGO was found to show the lowest T_{red}. Ud-din et. al [5] suggested that metals with smaller size are easier to be reduced (higher reducibility) compared to larger one. The increase in particle size could create mass transfer limitations for the diffusion of hydrogen inside nickel oxide particles, leading to the higher energy required to reduce the Ni^{2+} to Ni^0. This agrees with the XRD result from this experiment where the loading of copper resulted in a smaller crystallite size of the catalysts and a higher surface area. This is, again, correlated to the dispersion of metal on the support. As a result, a lower temperature for reduction of the catalyst can be used for the activation before carbon dioxide methanation activity.

4 Conclusion

This work presents the successful synthesis and characterization of Ni-M/rGO catalysts. The influence of second metals namely cobalt and copper were studied in terms of the effect towards the crystallinity, surface area, and reducibility of the catalyst. The addition of metal was found to shrink the size of the catalysts which is related to the dispersion on rGO support. As a result of this reduced size, the reducibility of the catalyst positively impacted where lower reduction temperature is achieved under the presence of second metal in the bimetallic catalyst system. Way forward, the performance of the bimetallic catalyst will be studied in terms of CO_2 conversion activity in comparison to Ni/rGO catalyst which shows 51% CO_2 conversion, as reported earlier [17].

Acknowledgements The financial assistance from the Foundation of Universiti Teknologi PETRONAS (YUTP-FRG) with the cost center 015LC0-253 is gratefully acknowledged.

References

1. M. Aziz, A. Jalil, S. Triwahyono, A. Ahmad, CO_2 methanation over heterogeneous catalysts: recent progress and future prospects. Green Chem. **17**, 2647–2663 (2015)
2. M. Cai, J. Wen, W. Chu, X. Cheng, Z. Li, Methanation of carbon dioxide on Ni/ZrO_2-Al2O_3 catalysts: effects of ZrO_2 promoter and preparation method of novel ZrO_2-Al2O_3 carrier. J. Nat. Gas Chem. **20**, 318–324 (2011)
3. F. Ocampo, B. Louis, L. Kiwi-Minsker, A.-C. Roger, Effect of Ce/Zr composition and noble metal promotion on nickel based CexZr$_1$−xO_2 catalysts for carbon dioxide methanation. Appl. Catal. A **392**, 36–44 (2011)
4. K. Stangeland, D. Kalai, H. Li, Z. Yu, CO_2 methanation: the effect of catalysts and reaction conditions. Energy Procedia **105**, 2022–2027 (2017)
5. I. Ud Din, M.S. Shaharun, D. Subbarao, A. Naeem, Synthesis, characterization and activity pattern of carbon nanofibers based copper/zirconia catalysts for carbon dioxide hydrogenation to methanol: influence of calcination temperature. J. Power Sourc. **274**, 619–628 (2015)
6. A. Aljishi, G. Veilleux, J.A.H. Lalinde, J. Kopyscinski, The effect of synthesis parameters on ordered mesoporous nickel alumina catalyst for CO_2 methanation. Appl. Catal. A **549**, 263–272 (2018)
7. M.A.A. Aziz, A.A. Jalil, S. Triwahyono, M.W.A. Saad, CO_2 methanation over Ni-promoted mesostructured silica nanoparticles: influence of Ni loading and water vapor on activity and response surface methodology studies. Chem. Eng. J. **260**, 757–764 (2015)
8. P. Frontera, A. Macario, M. Ferraro, P. Antonucci, Supported catalysts for CO_2 methanation: a review. Catalysts **7** (2017)
9. X. Su, J. Xu, B. Liang, H. Duan, B. Hou, Y. Huang, Catalytic carbon dioxide hydrogenation to methane: a review of recent studies. J Energy Chem **25**, 553–565 (2016)
10. J. Gao, Q. Liu, F. Gu, B. Liu, Z. Zhong, F. Su, Recent advances in methanation catalysts for the production of synthetic natural gas. RSC Adv. **5**, 22759–22776 (2015)
11. F. Meshkani, M. Rezaei, Nanocrystalline MgO supported nickel-based bimetallic catalysts for carbon dioxide reforming of methane. Int. J. Hydrogen Energy **35**, 10295–10301 (2010)
12. C.-J. Liu, J. Ye, J. Jiang, Y. Pan, Progresses in the preparation of coke resistant Ni-based catalyst for steam and CO_2 reforming of methane. Chem. Cat. Chem. **3**, 529–541 (2011)
13. S. Hwang, U.G. Hong, J. Lee, J.H. Baik, D.J. Koh, H. Lim, I.K. Song, Methanation of carbon dioxide over mesoporous nickel–M–alumina (M=Fe, Zr, Ni, Y, and Mg) xerogel catalysts: effect of second metal. Catal. Lett. **142**, 860–868 (2012)
14. J.-N. Park, E.W. McFarland, A highly dispersed Pd–Mg/SiO_2 catalyst active for methanation of CO_2. J. Catal. **266**, 92–97 (2009)
15. Y. Li, G. Lu, J. Ma, Highly active and stable nano NiO–MgO catalyst encapsulated by silica with a core–shell structure for CO_2 methanation. RSC Adv. **4**, 17420–17428 (2014)
16. G.B. Sommerbauer, S. Lux, M. Siebenhofer, Metahanation of carbon dioxide with a novel nickel/magnesium oxide catalyst, AIChE Annual MeetingSan Francisco, 13–18 Nov. 2016
17. N.D. Mohd Ridzuan, M.S. Shaharun, K.M. Lee, I. Ud Din, P. Puspitasari, Influence of nickel loading on reduced graphene oxide-based nickel catalysts for the hydrogenation of carbon dioxide to methane. Catalysts **10**, 471 (2020)
18. Z. Zhang, Y. Tian, L. Zhang, S. Hu, J. Xiang, Y. Wang, L. Xu, Q. Liu, S. Zhang, X. Hu, Impacts of nickel loading on properties, catalytic behaviors of Ni/γ–Al_2O_3 catalysts and the reaction intermediates formed in methanation of CO_2. Int. J. Hydrogen Energy **44**, 9291–9306 (2019)
19. J. Tan, J. Wang, Z. Zhang, Z. Ma, L. Wang, Y. Liu, Highly dispersed and stable Ni nanoparticles confined by MgO on ZrO_2 for CO_2 methanation. Appl. Surf. Sci. **481**, 1538–1548 (2019)
20. S. Lowell, J.E. Shields, M.A. Thomas, M. Thommes, *Characterization of Porous Solids and Powders: Surface Area, Pore Size and Density* (Springer Science & Business Media, 2012)
21. D. Ramimoghadam, M.Z.B. Hussein, Y.H. Taufiq-Yap, Hydrothermal synthesis of zinc oxide nanoparticles using rice as soft biotemplate. Chem. Cent. J. **7**, 136 (2013)
22. W. Gac, W. Zawadzki, G. Słowik, A. Sienkiewicz, A. Kierys, Nickel catalysts supported on silica microspheres for CO_2 methanation. Microporous Mesoporous Mater. **272**, 79–91 (2018)

23. X. Jia, X. Zhang, N. Rui, X. Hu, C.-J. Liu, Structural effect of Ni/ZrO_2 catalyst on CO_2 methanation with enhanced activity. Appl. Catal. B **244**, 159–169 (2019)
24. M. Romero-Sáez, A.B. Dongil, N. Benito, R. Espinoza-González, N. Escalona, F. Gracia, CO_2 methanation over nickel-ZrO_2 catalyst supported on carbon nanotubes: a comparison between two impregnation strategies. Appl. Catal. B **237**, 817–825 (2018)
25. R. Daroughegi, F. Meshkani, M. Rezaei, Enhanced activity of CO_2 methanation over mesoporous nanocrystalline Ni–Al_2O_3 catalysts prepared by ultrasound-assisted co-precipitation method. Int. J. Hydrogen Energy **42**, 15115–15125 (2017)
26. S.V. Moghaddam, M. Rezaei, F. Meshkani, R. Daroughegi, Synthesis of nanocrystalline mesoporous Ni/$Al_2O_3SiO_2$ catalysts for CO_2 methanation reaction. Int. J. Hydrogen Energy **43**, 19038–19046 (2018)
27. J. Lin, C. Ma, Q. Wang, Y. Xu, G. Ma, J. Wang, H. Wang, C. Dong, C. Zhang, M. Ding, Enhanced low-temperature performance of CO_2 methanation over mesoporous Ni/Al_2O_3-ZrO_2 catalysts. Appl. Catal. B **243**, 262–272 (2019)
28. M. Mihet, M.D. Lazar, Methanation of CO_2 on Ni/γ-Al_2O_3: influence of Pt, Pd or Rh promotion. Catal. Today **306**, 294–299 (2018)
29. L. Hu, A. Urakawa, Continuous CO_2 capture and reduction in one process: CO_2 methanation over unpromoted and promoted Ni/ZrO_2. J. CO_2 Utilization **25**, 323–329 (2018)

Photocatalytic Pre-treatment of Lignocellulosic Biomass for Biohydrogen Production

Nurul Tasnim Sahrin, Jun Wei Lim, Fatima Musa Ardo, and Rashid Shamsuddin

Abstract Production of hydrogen utilizing lignocellulosic biomass from microalgae is one of the most demanding technologies for sustainably generating energy considering environmental concerns. Nevertheless, an efficient conversion process is still difficult to be achieved due to the complex nature of biomass. In this study, a simple and mild process known as the photocatalytic pre-treatment process employing TiO_2 nanotubes as a photocatalyst was used to disrupt the lignocellulose complex. As compared to raw palm kernel expeller, maximum soluble carbohydrate conversion was achieved at 9%. Pre-treatment of palm kernel expeller + light and palm kernel expeller + TiO_2 nanotubes successfully increased the soluble carbohydrate conversion to 18% and 24%, respectively. However, a significant enhancement of maximum soluble carbohydrate conversion was observed for palm kernel expeller + Light + TiO_2 nanotubes photocatalytic pre-treatment which was 40%. The hydrogen yield for raw palm kernel expeller, palm kernel expeller + Light, palm kernel expeller + TiO_2, and palm kernel expeller + Light + TiO_2 were 290, 340, 360, and 430 mL/g S_{carb}, respectively. Comparatively, a high enhancement of hydrogen yield was observed due to increased soluble carbohydrate conversion.

Keywords Hydrogen · TiO_2 photocatalyst · Lignocellulosic biomass

N. T. Sahrin · J. W. Lim (✉) · F. M. Ardo
HICoE-Centre for Biofuel and Biochemical Research (CBBR), Institute of Self-Sustainable Building (ISB), Department of Fundamental and Applied Sciences, Universiti Teknologi PETRONAS, 32610 Seri Iskandar, Perak Darul Ridzuan, Malaysia
e-mail: junwei.lim@utp.edu.my

N. T. Sahrin
e-mail: nurul_22005153@utp.edu.my

F. M. Ardo
e-mail: fatima_22000759@utp.edu.my

R. Shamsuddin
HICoE-Centre for Biofuel and Biochemical Research (CBBR), Institute of Self-Sustainable Building (ISB), Department of Chemical Engineering, Universiti Teknologi PETRONAS, 32610 Seri Iskandar, Perak Darul Ridzuan, Malaysia
e-mail: mrashids@utp.edu.my

M. B. Othman et al. (eds.), *Proceedings of the 1st International Conference of New Energy*, Springer Proceedings in Energy, https://doi.org/10.1007/978-981-99-0859-2_7

1 Introduction

Sustainability has gained traction in the face of the current worldwide phenomenon of rising energy demands paired with environmental deterioration, igniting interest in biofuels as alternative and complementary bioenergy sources [1]. Fossil fuels contribute to the largest percentage of the world energy matrix, and these fuels are depletable as well as responsible for the release of gases contributing to the greenhouse effect. Owing to the rising atmospheric carbon dioxide (CO_2) concentration, hydrogen could play an important role in the global energy transformation to meet the energy requirements, and its consumption as fuel is completely deprived of CO_2 emissions [2]. As microalgae have been extensively investigated to make bioethanol and biodiesel, they are thought to be a high-yield source for hydrogen generation. Besides, due to their higher rate of photosynthesis and faster pace of growth than other terrestrial plants, microalgae have also attracted a lot of interest as a substitute feedstock for biofuels.

Indeed, microalgae are photosynthesis microorganisms that utilize solar energy to create biologically significant products such as lipids, proteins, and carbohydrates [3]. According to previous research, high carbohydrate content can contribute to the likelihood of its conversion into biohydrogen [4]. Carbohydrates are found primarily as a polysaccharide that is produced and stored in microalgal cells in the form of starch [1]. Carbohydrates are then consumed by metabolic activities, creating excessive electrons for disposal [5]. Subsequently, the electron transport chain will function when there is no dissolved O_2 present in the medium. The O_2 depletion will create an anaerobic condition in the culture which causes the activation of O_2 sensitive hydrogenase enzyme, and in turn, produces hydrogen from the electrons generated [2].

Besides, it is crucial to note that the choice of organic carbon sources is crucial for both carbohydrate buildup and cost-effectiveness as nutrient account for over 25% of overall production expenditures [6]. Therefore, the exploration of cost-effective techniques has thus turned into a crucial component of biohydrogen generation to ensure sustainable output. Palm kernel expeller (PKE) is a solid waste produced from the palm milling industry and can act as a replacement for liquid carbon sources. Lignin, hemicellulose, and cellulose are the main components of lignocellulosic biomass. However, lignin is regarded as a component that has inhibitory effects on lignocellulosic biomass conversion [7].

Therefore, numerous studies have investigated lignin degradation to break the rigid lignin structure to ease the accessibility of cellulose. There are various pretreatment methods that have been adopted including pyrolysis, catalytic, enzymatic, and ozonolysis. Among them, photocatalytic degradation is considered the most feasible and environmentally friendly approach for lignin degradation [8]. In this research, TiO_2 nanotube photocatalyst was utilized for lignin degradation due to its enhanced photoactivity under visible light.

2 Materials and Methodology

2.1 *Synthesis of TiO_2 Nanotubes Photocatalyst*

Glycerol-choline chloride electrolyte was firstly synthesized by mixing choline chloride (98% ChCl, Sigma Aldrich) with glycerol (98% $C_3H_8O_3$, Fischer Scientific) at a molar ratio of 1:2 and was directly heated for 30 min at 80 °C to form a clear solution [9]. The electrolyte was then used for the anodization process in the synthesis of TiO_2 nanotubes. In brief, the anodization setup comprised of the platinum rod as cathode and the anode was made up of technical grade Ti foil with a thickness of 0.1 mm and was cut into 2 cm × 1 cm (Titanium, Ti Gr5/Tc4 Grade 5 ASTM B265 Thin Plate Sheet). The experiment was carried out at an anodization voltage of 30 V for 1 h under room temperature. After the completed anodization experiment, the synthesized sample was separated instantly from the electrolyte solution and rinsed with distilled water. Subsequently, the sample was air dried under room temperature and calcined at 300 °C for 2 h.

2.2 *Photocatalytic Pre-treatment of Palm Kernel Expeller*

TiO_2 nanotubes photocatalyst was dipped into a 500 mL solution consisting of 0.5 g of PKE. The solution was magnetically stirred at 400 rpm and kept under visible light using a 500 W halogen lamp for 3 h. Liquid samples were taken at a regular interval of 1 h for a total of 3 h. After pre-treatment, the liquid samples were filtered through 0.22 μm syringe filter prior soluble carbohydrate and soluble carbon oxygen demand (COD) analysis. Soluble carbohydrate was measured following the phenol–sulfuric method [10]. Meanwhile, the soluble COD analysis was analyzed according to the Standard Methods APHA5520C [11].

2.3 Chlorella Vulgaris *Stock Cultivation*

The freshwater microalgae species, *Chlorella vulgaris,* was obtained from the culture collections owned by the Center of Biofuel and Biochemical Research (CBBR), Universiti Teknologi PETRONAS. The 500 mL of microalgae seed was grown in a 5 L glass reactor at a temperature of 25 °C with an ambient air aeration rate of 6.5 L/min and lighting from a cool-white fluorescent lamp at a light intensity of 60–70 $\mu mol/m^2s$. Throughout the cultivation phase, a medium consisting of K_2HPO_4 (0.3375 g/L), KH_2PO_4 (0.7875 g/L), and $NaNO_3$ (0.25 g/L) was used and the pH was kept at 7.0 ± 0.1. Before being employed in the subsequent experimental setup, the microalgae were cultured until they reached the stationary growth phase.

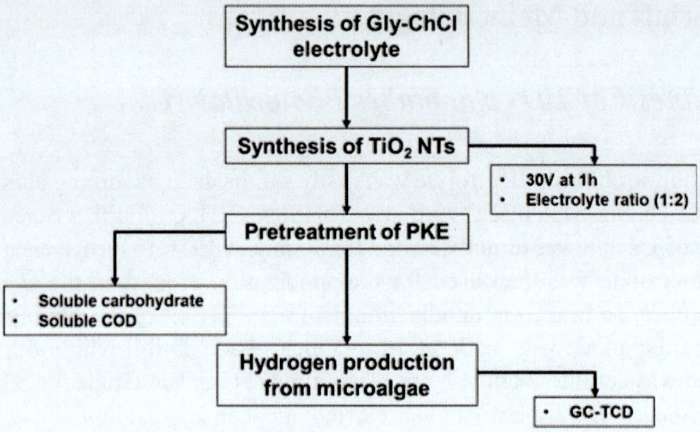

Fig. 1 Summary of overall experimental set up

2.4 *Experimental Setup for Hydrogen Production Using Palm Kernel Expeller*

The organic carbon source, PKE, was collected from the local producer and stored at room temperature. A batch test was conducted in 500 mL Schott bottles containing 1 g/L of PKE and 0.25 g/L of microalgae with a total working volume of 500 mL. Subsequently, the bottles were flushed with nitrogen gas for 20 min to achieve anaerobic conditions and the cap was closed directly. The mixtures were magnetically stirred at 80 rpm/min at room temperature and each test was conducted in duplicate. The hydrogen production was measured by the withdrawal of the gas produced at the headspace of the Schott bottle using a gas-tight syringe. 1 mL of gas sample was withdrawn and injected into the chromatography-thermal conductivity detector (GC-TCD) using argon gas as a carrier gas. The concentration of hydrogen gas detected in the gas sample was multiplied by the total headspace volume in the bottle to obtain the total hydrogen production (Fig. 1).

3 Results and Discussion

Experiments for assessing photocatalytic treatment of PKE by TiO_2 nanotubes was conducted to evaluate the effect of different pre-treatment conditions on soluble carbohydrate as shown in Fig. 2a. As illustrated in Fig. 2a, raw PKE without pre-treatment act as a control in this study which was left at room temperature has shown no significant improvement in soluble carbohydrate conversion. However, there was a steady increase with approximately 40% of soluble carbohydrate conversion was achieved for photocatalytic pre-treatment of PKE by TiO_2 nanotubes under visible light illumination after 3 h of pre-treatment duration. Meanwhile, pre-treatment conditions for PKE + TiO_2 and PKE + Light have shown a plateau after 2 h with 23% and 18% of soluble carbohydrate conversion efficiencies, respectively. The results clearly indicated that the pre-treatment with the TiO_2 nanotubes photocatalyst under visible light irradiation has a significant effect on the degradation of lignocellulosic biomass in comparison with the pre-treatment in the presence of visible light and photocatalyst only.

This phenomenon is possibly due to the photocatalyst's ability to generate a large amount of hydroxyl radicals (·OH) and superoxide radicals ($\cdot O_2^-$) on its surface. The ·OH will react with the biomass molecules and result in lignin degradation due to the scission of β-O-4 bond which is accountable for 35–50% by weight of lignin [12]. Besides, ·OH can also directly attack the phenyl rings of lignin [13]. High soluble carbohydrate content is desirable as carbohydrates are recognized as a preferable substrate for hydrogen production [14]. In addition, soluble COD was observed to increase as pre-treatment duration increased to 3 h for PKE + Light + TiO_2 as shown in Fig. 2b. It can be inferred that with an increase in pre-treatment duration, more dissolved organic matter is available. Theoretically, highly dissolved organic materials are favorable to produce highly renewable hydrogen [15].

For evaluating the effect of different pre-treatment conditions, a series of experiments on hydrogen production from microalgae was conducted as shown in Fig. 3a. Raw PKE (control) exhibited a hydrogen yield of 290 mL/g S_{carb}. Meanwhile, samples for PKE + Light, PKE + TiO_2, and PKE + Light + TiO_2, display a hydrogen yield of 340, 360, and 440 mL/g S_{carb}, respectively. A significant enhancement of hydrogen yield was observed for PKE + Light + TiO_2, the probable reason might be due to the exposure to high availability of cellulose and hemicellulose contents resulting from delignification and partial solubilization of hemicellulose due to the pre-treatment process [12].

The stability of the photocatalyst is a crucial consideration for an industrial-scale application. In order to illustrate the stability of the synthesized TiO_2 photocatalyst, the photocatalyst was reused for the pre-treatment study for three consecutive cycles. The results of the recyclability study on hydrogen yield are displayed in Fig. 3b. The

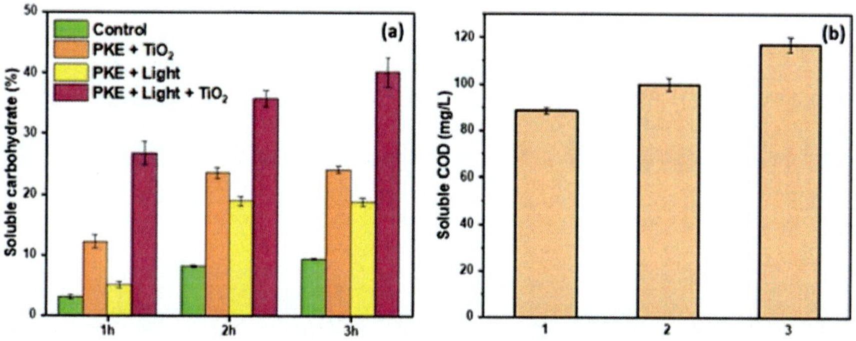

Fig. 2 **a** Soluble carbohydrate conversion efficiencies (%) and **b** Soluble COD concentration (mg/L)

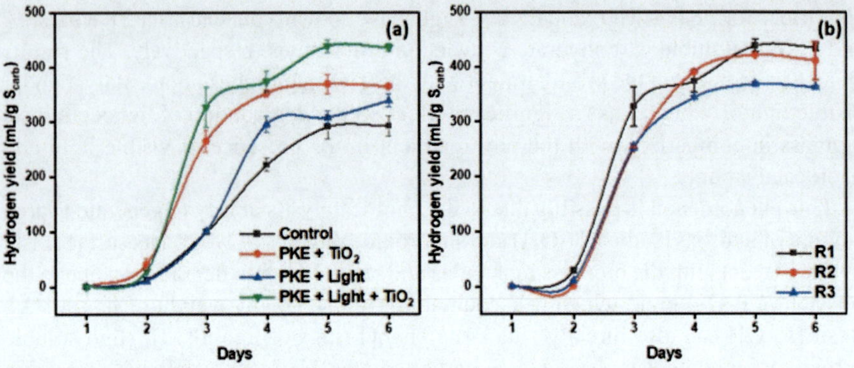

Fig. 3 **a** Hydrogen yield (mL/g S_{carb}) at different pre-treatment conditions and **b** hydrogen yield (mL/g S_{carb}) for three consecutive cycles

hydrogen yield was reduced from 440 mL/g S_{carb} to 350 mL/g S_{carb}. The results indicated that the synthesized photocatalyst is stable for the conversion of soluble carbohydrates for subsequent hydrogen production.

4 Conclusions

Photocatalytic pre-treatment of palm kernel expeller using TiO_2 photocatalyst was observed to show an exemplary performance for high soluble carbohydrate conversion and high dissolved organic matter available for subsequent hydrogen production under 3 h of visible light irradiation. The improved soluble carbohydrate and dissolved organic matter show an enhancement of hydrogen production from 290 mL/g S_{carb} to 440 mL/g S_{carb}. Moreover, the TiO_2 photocatalyst exhibited good stability for three consecutive cycles for soluble carbohydrate conversion in producing hydrogen from microalgae.

Acknowledgements The authors would like to acknowledge the funding support from Ministry of Higher Education (MOHE) Malaysia through HICoE Grant (cost center 015MA0-052/015MA0-104/015MA0-136) to CBBR. Yayasan Universiti Teknologi PETRONAS via YUTP-FRG with the cost center of 015LC0-341 is also gratefully acknowledged.

References

1. M. de Carvalho Silvello et al., Microalgae-based carbohydrates: a green innovative source of bioenergy. Bioresour. Technol. **344**, 126304 (2021)
2. A. Hemschemeier, A. Melis, T. Happe, Analytical approaches to photobiological hydrogen production in unicellular green algae. Photosynth. Res. **102**(2), 523–540 (2009)

3. R. Maurya et al., Advances in microalgal research for valorization of industrial wastewater. Bioresour. Technol. **343**, 126128 (2022)
4. G.E. Lakatos, K. Ranglová, J.C. Manoel, T. Grivalský, J. Kopecký, J. Masojídek, Bioethanol production from microalgae polysaccharides. Folia Microbiol. (Praha) **64**(5), 627–644 (2019)
5. A. Melis, L. Zhang, M. Forestier, M.L. Ghirardi, M. Seibert, Sustained photobiological hydrogen gas production upon reversible inactivation of oxygen evolution in the green alga Chlamydomonas reinhardtii. Plant Physiol. **122**(1), 127–136 (2000)
6. J.A.V. Costa, B.C.B. Freitas, G.M. Rosa, L. Moraes, M.G. Morais, B.G. Mitchell, Operational and economic aspects of Spirulina-based biorefinery. Bioresour. Technol. **292**, 121946 (2019)
7. M. Mahdavi, M. Mirmohammadi, M. Baghdadi, S. Mahpishanian, Visible light photocatalytic degradation and pretreatment of lignin using magnetic graphitic carbon nitride for enhancing methane production in anaerobic digestion. Fuel **318**, 123600 (2022)
8. R. Sankaran et al., Recent advances in the pretreatment of microalgal and lignocellulosic biomass: a comprehensive review. Bioresour. Technol. **298**, 122476 (2020)
9. D.P. Jaihindh, Y.P. Fu, Facile synthesis of deep eutectic solvent assisted BiOCl/$BiVO_4$@AgNWs plasmonic photocatalysts under visible light enhanced catalytic performance. Catal. Today **297**, 246–254 (2017)
10. D. Herbert, P.J. Phipps, R.E. Strange, Chapter III chemical analysis of microbial cells. Methods Microbiol. **5**, 209–344 (1971)
11. K. Hua Ang, Potentially land used pattern contributing pollution source towards water quality: a case study of UTM river **1**, 1–4 (2017)
12. M. Sabeeh, Zeshan, R. Liaquat, A. Maryam, Effect of alkaline and alkaline-photocatalytic pretreatment on characteristics and biogas production of rice straw. Bioresour. Techno. **309**, 12344 (2020)
13. X. Liu, X. Duan, W. Wei, S. Wang, B.J. Ni, Photocatalytic conversion of lignocellulosic biomass to valuable products. Green Chem. **21**(16), 4266–4289 (2019)
14. H. Zhu et al., Biohydrogen production by anaerobic co-digestion of municipal food waste and sewage sludges. Int. J. Hydrog. Energy **33**(14), 3651–3659 (2008)
15. R.A. Rozendal, H.V.M. Hamelers, G.J.W. Euverink, S.J. Metz, C.J.N. Buisman, Principle and perspectives of hydrogen production through biocatalyzed electrolysis. Int. J. Hydrog. Energy **31**(12), 1632–1640 (2006)

Effect of Nitrogen Doping on Optoelectronic Properties of TiO_2 Anatase Model for Solar Hydrogen Production: A DFT + *U* Approach

Farman Ullah, Beh Hoe Guan, Siti Nur Azella Zaine, Usman Ghani, and Mohamed Shuaib Mohamed Saheed

Abstract To deal with the energy crisis and environmental challenges, solar hydrogen generation via photocatalytic water-splitting technique is clean and green energy technology. Titanium dioxide (TiO_2) plays a significant role as a photocatalyst to absorb solar energy for photocatalytic H_2 production. However, the development of TiO_2 as an efficient photocatalyst is always a challenging task due to its wide bandgap (TiO_2 anatase ~3.2 eV) and meager visible light absorption. Herein, this work presents the computationally designed nitrogen (N)-doped TiO_2 anatase models

F. Ullah · B. H. Guan
Department of Fundamental and Applied Sciences, Universiti Teknologi PETRONAS, 32610 Seri Iskandar, Perak, Malaysia
e-mail: farman_19001014@utp.edu.my

B. H. Guan
e-mail: beh.hoeguan@utp.edu.my

F. Ullah · S. N. A. Zaine · M. S. M. Saheed (✉)
Centre of Innovative Nanostructure and Nanodevices (COINN), Institute of Autonomous System, Universiti Teknologi PETRONAS, 32610 Seri Iskandar, Perak, Malaysia
e-mail: shuaib.saheed@utp.edu.my

S. N. A. Zaine
e-mail: nurazella.zaine@utp.edu.my

F. Ullah
Department of Physics, University of Science and Technology, Bannu 28100, Khyber Pakhtunkhwa, Pakistan

S. N. A. Zaine
Department of Chemical Engineering, Universiti Teknologi PETRONAS, Seri Iskandar, Malaysia

U. Ghani
State Key Laboratory of Metal Matrix Composites, Shanghai Jiao Tong University, Shanghai 200240, PR China

M. S. M. Saheed
Department of Mechanical Engineering, Universiti Teknologi PETRONAS, 32610 Seri Iskandar, Perak, Malaysia

M. B. Othman et al. (eds.), *Proceedings of the 1st International Conference of New Energy*, Springer Proceedings in Energy, https://doi.org/10.1007/978-981-99-0859-2_8

simulated via periodic density functional theory (DFT) calculations over large supercells. Hubbard's modified DFT calculations were adopted through Perdew–Burke–Ernzerhof supported generalized gradient approximation (GGA + PBE + U) functional to simulate the optoelectronic properties of the designed models. The results reveal that N-doped TiO_2 anatase model exhibits a substantial bandgap reduction up to 2.34 eV as endorsed by the electronic structure analysis. The bandgap reduction commences from the provision of N 2p states to the O 2p and Ti 3d states of TiO_2 in VB region and their presence as induced mid-gap states in the bandgap. The reduction in bandgap energy of the TiO_2 significantly boosts the visible light absorption under solar irradiation. Thus overall, the N-doping could be a promising non-metal doping approach for TiO_2 anatase photocatalyst for the solar H_2 production process.

Keywords Non-metal doping · Density functional theory · Bandgap energy · Density of states · Optical absorption

1 Introduction

The most imperative challenge for the scientific community is to develop a long-term clean energy economy to excellently balance the increasing demand for energy and the threatening effect of global warming [1, 2]. Solar Hydrogen (H_2) energy, which can be created through the electrolysis of water via solar energy utilization is one of the most promising choices as a solar fuel and is believed to be a sound alternative from a future perspective. The development of an efficient semiconductor photocatalyst will play a dominant role in solar H_2 production through the photocatalytic water-splitting technique. Titanium dioxide (TiO_2) is considered an auspicious option among different semiconductor photocatalysts due to its excellent photo and electrochemical stability, good photocatalytic activity (high reaction rates), suitable band edge position for H_2 evolution reaction, low production cost, large natural abundance, operation under ambient conditions of temperature and pressure, and non-toxicity [3, 4]. Regardless of the above-mentioned merits, deprived absorbance in the visible spectral region and faster recombination of photogenerated electron–hole (e^-–h^+) pairs are still considered the two predominant drawbacks that lead to the low quantum efficiency and hinder the developments of TiO_2 nanostructure photocatalyst. Various structure modifications techniques and synthetic strategies including metal and non-metal doping of TiO_2 are explored to surmount these two obstacles and reinforce the photoactivities of TiO_2 photocatalyst [5].

Non-metal doping commences impurity energy states into the bandgap structure to shift the CBM positions to decrease the bandgap and thus, increase utilization of the visible irradiation from the solar spectra positions [7–9]. Bandgap reduction and superior absorption of visible light from solar energy will result in enhanced photocatalytic performance [6, 7]. Many researchers claimed that effective TiO_2 doping may alter the response of TiO_2 to move into the visible photo-spectra. Batalović et al.

[5] reported N Doping of TiO_2 anatase photocatalysts via theoretical DFT calculations and reported enhanced photoactivity for the N-doped TiO_2 anatase models. Furthermore, they reported that the photocatalytic properties greatly depend upon the synthesis procedure. Chen et al. [8] assessed the influence of nitrogen and sulfur (N, S) dopants through interstitial and substitutional doping approaches for the TiO_2 model and argued that the designed interstitial doped TiO_2 model carries higher absorption ability in comparison to the substitutional doped TiO_2 models. Although, non-metal doping had a considerable impact on the performance of TiO_2. However, it is still difficult to determine the precise mechanism of doping because of the various preparation procedures and experimental conditions. Recent research demonstrated that TiO_2 anatase may be converted into a superior photocatalyst by adding an appropriate quantity of N-dopants into TiO_2 to enhance their photocatalytic performance.

Therefore, in this work, we study the influence of N dopants on the structural, and optoelectronic properties of the TiO_2 anatase model with Hubbard's mounted DFT simulation approach. For the first time, we examined the effect of N-dopants upon the TiO_2 anatase photocatalytic activity by substituting host O atoms with N atoms via the DFT + U technique. The electronic structure simulation results exhibit that the N-doped TiO_2 anatase model endorsed a noteworthy bandgap reduction by comparing with the undoped TiO_2 anatase model. The bandgap value for the designed N-doped TiO_2 anatase models was reduced to 2.34 eV. The reduced bandgap signifies the visible absorbance of the N-doped TiO_2 model up to 438 nm in visible spectra to improve the photoactivity accordingly.

2 Methodology

The present study employed theoretical calculations with the execution of the Cambridge sequential total energy package (CASTEP) code, established upon the pseudopotential approach adopted for the valence electron orbitals [9]. Hubbard's modified GGA + PBE + U functional was exercised to characterize the optoelectronic characteristics. The geometry optimization was done through a minimization algorithm via Broyden–Fletcher–Goldfarb–Shanno scheme [10, 11]. To effectively display a bulk solid-like behavior, the unit cell with optimized geometry was converted into a supercell. Also, the supercell approach was exercised to compensate for a vacancy/defect or if the designed system requires to convert into a doped system with the insertion of the dopant atoms.

Here, a DFT simulated $2 \times 2 \times 1$ supercell was introduced to develop the N-doped TiO_2 anatase model. For the N-doped TiO_2 anatase models, one, two, and three O atoms of the undoped TiO_2 anatase model (as reported previously [12]) were replaced by each N atom to have different N concentrations for the N-doped TiO_2 anatase models. The undoped TiO_2 anatase supercell contains a total of 48 atoms with 32 Ti atoms, 16 O atoms. For the N-doped TiO_2 anatase models, O atoms were replaced by corresponding N atoms in an undoped TiO_2 anatase system with 48 Ti atoms, 15 O atoms, and 1 N atom (Fig. 1). The optimized structure parameters were adjusted with the cutoff energy value of 400 eV, with the value of residual force ~0.02 eV/Å, atomic displacement ~0.001 Å, stress value ~0.05 GPa and energy values for convergence standards about ~1×10^{-6} eV. The $4 \times 4 \times 4$ Monkhorst–Pack k-point grid integration for the TiO_2 anatase unit cell was specified to reduce the calculation time. The configurations of valence electronic orbits were contemplated with Ti ($3s^2 3p^6 3d^2 4s^2$), O ($2s^2 2p^4$), and N ($2s^2 2p^3$).

Although, standard DFT calculations estimate the optoelectronic behavior of the theoretically designed computational models very well and predict more comparable qualities to the experimental observation. However, for transition metal oxide strongly correlated TiO_2-based systems [13, 14], the standard DFT calculations underestimate the bandgap energy calculations. Therefore, Hubbard's onsite potential is mostly preferred for the computation of electronic structure and related characteristics due to comparatively more accurate simulation results at less computational

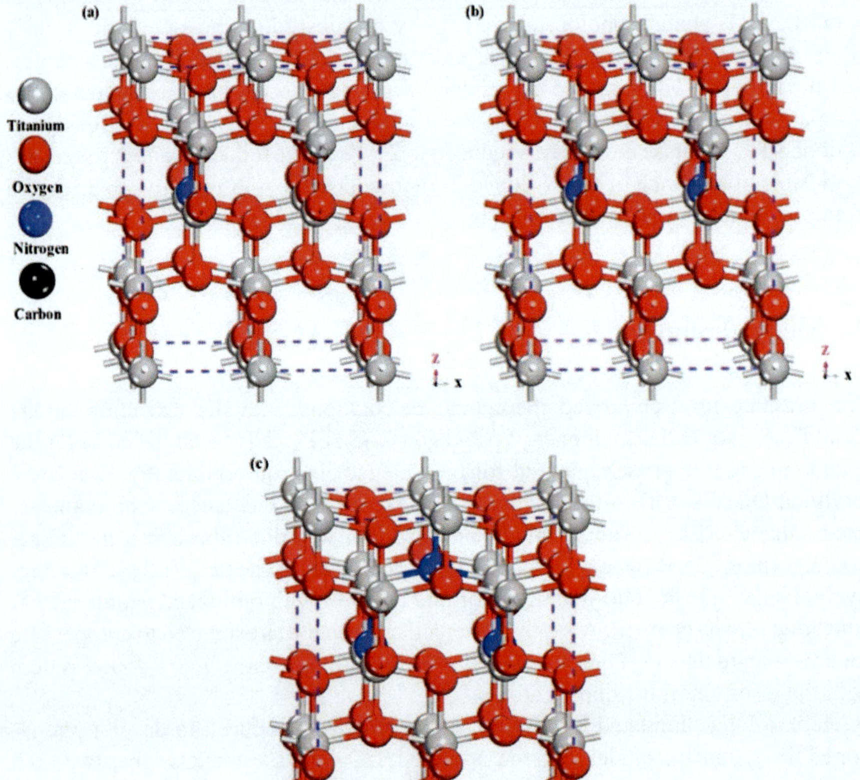

Fig. 1 N-doped TiO_2 anatase computational models simulated via DFT + U technique with N concentrations **a** 2.08 at.%, **b** 4.17 at.%, and **c** 6.25 at.%, simulated via DFT simulation

cost [15, 16]. In this technique, Hubbard potential is utilized to acquire more accurate results. Various Hubbard parameter (U) values were adopted for the valence Ti 3d electrons to have the bandgap energy accordingly. Finally, the U value of 8.47 eV was employed for the strongly localized electrons and thus the obtained simulation results approach experimental results very well.

3 Results and Discussion

3.1 *Structural Properties and Stability of the Models*

The computed lattice parameters and average bond lengths of the geometrically optimized N-doped TiO_2 supercell models are tabulated below in Table 1. The optimized lattice parameters estimated for undoped TiO_2 are a = b = 3.797 Å and c = 9.556 Å, as reported in our previous work [12], and are in decent agreement with the reported experimental values of a = b = 3.782 Å, c = 9.502 Å [17]. For an undoped TiO_2 anatase model, each Ti atom is linked to four nearest and two 2nd nearest O neighbors. Therefore, the average bond lengths are generally characterized as the nearest Ti–O bond and 2nd nearest Ti–O bonds. For undoped TiO_2 anatase, the average bond length of the nearest Ti–O bond is 1.947 Å, whereas for the 2nd nearest Ti–O bond, the bond length is ~1.983 Å, as reported previously in our work [12]. In contrast, for the N-doped TiO_2 anatase models, the Ti–O bond lengths appeared to be significantly longer than those of the undoped TiO_2 anatase models. By increasing the N dopants concentration, the Ti-N bond lengths also tend to increase. Due to the fact that the majority of the Ti–N bond lengths are significantly longer than those of the T–O bonds at the same nitrogen concentration. Therefore, an increase in N-concentration tends to be accompanied by an increase in the volume ratio of the N-doped TiO_2 (V_N) in contrast to the volume ratio of the undoped TiO_2 (V_p). The results signify that N-doping roots an expansion in the volume of the unit cell due to the following two reasons. The first one is the difference in radii of the two ions. For example, for the N^{3-} ions, its value is ~2.032 Å and for the O^{2-} ions, it is around 2.260 Å. Second reason is related to the electronegativity of the two atoms. As the N atom has a lower electronegativity than the O atom, the attraction between the two atoms in the Ti–N bond is reduced, resulting in a bond with longer bond length.

To assess the relative stability of the different N-doped TiO_2 anatase models, the defect formation energies were computed as defined in the below equation:

$$E_f = E_{\text{tot}}(\text{N - doped}) - E_{\text{tot}}(\text{undoped}) - n\mu_{\text{N}} - n\mu_{\text{O}} \quad (1)$$

Here, E_{tot}(undoped) and E_{tot}(N-doped) correspond to the total energies of the undoped and N-doped TiO_2 anatase models, respectively, n denotes the number of N substituted atoms, and μ_N and μ_O correspond to the chemical potentials of N and O atoms, respectively.

Table 1 Simulated lattice parameters and bond lengths of the geometrically optimized N-doped TiO_2 supercell models

N-doped TiO_2 Models	Cell lattice parameters			Bond lengths (Å)				Volume ratio
	a (Å)	b (Å)	c (Å)	Nearest Ti–O	2nd Nearest Ti–O	Nearest Ti–N	2nd Nearest Ti–N	ΔV $[V_N/V_{Un}]$ ($Å^3$)
Undoped TiO_2 [12]	3.797	3.797	9.556	1.947	1.983	–	–	1.0000
1N-doped TiO_2	3.771	3.792	9.560	1.931	1.992	1.934	2.022	1.0035
2N-doped TiO_2	3.766	3.825	9.519	1.934	1.996	2.032	2.260	1.0064
3N-doped TiO_2	3.815	3.799	9.613	1.945	1.988	2.030	2.263	1.0212

The formation energy is influenced by the growth circumstances, which can either be Ti-rich or O-rich. It is important to highlight that for the smaller E_f value, it is easier to incorporate the dopant atoms into the TiO_2 lattice. For the N-doped TiO_2 anatase models, the formation energies associated with Ti-rich conditions are lower than that associated with the O-rich conditions. It follows that the introduction of the N dopants at the O atom site into TiO_2 is more favorable than that at the Ti atom, which executes the incorporation of the N dopants into O matrix easily. Furthermore, by increasing the N dopants concentration, the formation energies are also increasing under both the Ti-rich and the O-rich circumstances. This suggests that at higher doping concentrations, the synthesis of the N-doped anatase TiO_2 suits is more challenging in comparison to lower N-doping concentrations.

3.2 *Electronic Structure/Characteristics*

To explore the electronic characteristics of the N-doped TiO_2 anatase model, the computed band structure projection for the N-doped TiO_2 anatase model is presented in Fig. 2. The structural changes due to the inclusion of the N dopants into the TiO_2 anatase model significantly change the electronic structures of the designed models. The Fermi energy level is taken as the zero-point energy. As for the undoped TiO_2 anatase model, the computed bandgap energy was 3.13 eV with a U parametric value of 8.47 eV, as reported in our previous work [12]. For N-doped TiO_2 anatase model, the bandgap energy is valued at ~2.34 eV. The value of the bandgap energy is significantly reduced than that of the undoped TiO_2 anatase model, as reported previously [12]. The plot display that the electronic structure of the undoped TiO_2 anatase model was altered significantly with the incorporation of N impurities, which further intensified the visible light absorption spectra too. The increase in the optical

Fig. 2 **a** Simulated electronic band structure of the N-doped TiO_2 anatase model

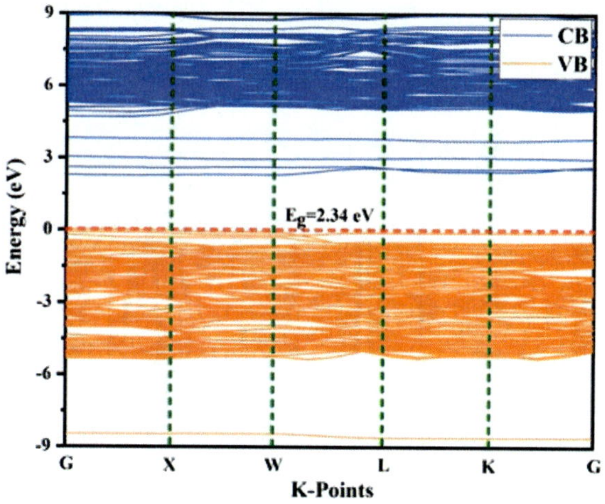

absorption peak intensity demonstrates the absorbance of more photons and the resultant redshift of the N-doped model appears more obvious. The enhanced visible light absorbance thus comparatively with less energetic visible light photons might excite more electrons too [18, 19]. The brilliant electronic characteristics of the N-doped TiO_2 anatase model is more promising for decreasing the electron transition energy and to improve the photocatalytic activity.

To examine further the composition of each of the associated orbitals, the partial density of states (PDOS) along with the total density of states (TDOS) analysis of the N-doped TiO_2 system was computed in Fig. 3. The bandgap narrowing due to the N-doping can be observed as the valence band maxima (VBM) is increased and conduction band minima (CBM) is decreased. As a result of the bandgap reduction, the required amount of electron transition energy to excite the electrons from VB to CB is decreased by approximately 0.79 eV and thus provokes a red shift of the absorbance edge. Furthermore, the incorporation of N-dopants into the TiO_2 broadens the VB and therefore heightens the mobility of the photogenerated e^-–h^+ pair. For the undoped TiO_2 anatase model, the VB largely comprises O 2p states along a minor contribution of Ti 3d states while the CB encompasses Ti 3d orbitals as well as with a smaller contribution of O 2p states which signifies that slight covalence bond characteristics can exist between Ti and O atoms. For the N-doped TiO_2 anatase model, the N atom may gain two electrons and consequently its 2p states are not completely occupied. This will lead to the appearance of shallow acceptor states located just over the VB states. Thus, the electrons in VB may first excite to the localized dopant states within the bandgap and successively to the CB after visible light photon absorption. However, it is important to highlight that the N 2p empty states may also act as trap centers and thus increasing the e^-–h^+ pair recombination. The full overlapping of the dopant impurity bands with VB can suppress the charge recombination to diminish the number of charge carrier traps. The CB steps toward the Fermi energy level which thus reduces the bandgap energy. The bandgap reduction

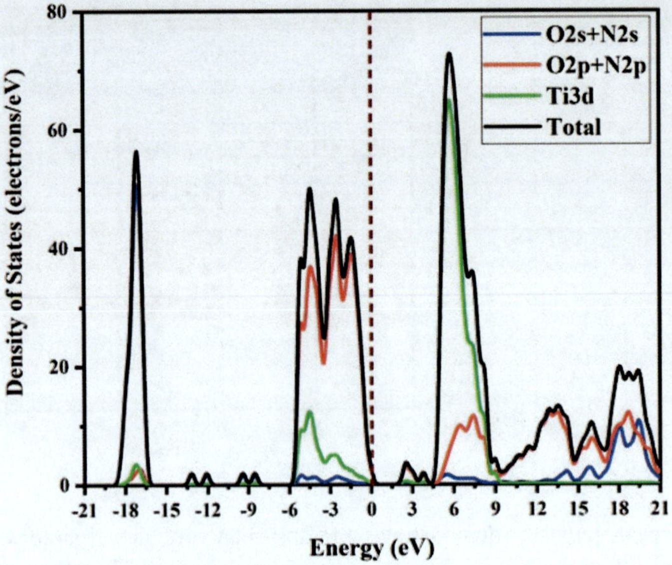

Fig. 3 The PDOS plots of the N-doped TiO_2 anatase model

in N-doped TiO_2 appears because of the conversion of Ti^{4+} to Ti^{3+} (reduction of Ti^{4+} to Ti^{3+}) and is due to the imbalance of charges between O^{2-} and N^{3-} ions [20].

3.3 Optical Properties

The optical properties were simulated via computation of the dielectric function via CASTEP code. The dielectric function comprises the real and imaginary parts with the real part acquired from Kramers–Kronig transforms whereas the imaginary part is calculated directly from the many-electron wave functions [21, 22]. The imaginary component of the dielectric function was regarded as the pandect of the optical properties as the inter-band optical transition between VB and CB can be studied by looking into it. Figure 4 demonstrates the simulated imaginary part of the dielectric functions $\varepsilon 2(\omega)$ of the N-doped TiO_2 anatase model, a large peak at 5.43 eV appeared due to the higher-energy transitions between parallel bands of the VB and CB. The major peak of the simulated functions was shifted downward toward the low-energy region. This shift may be due to the bandgap narrowing through the addition of N dopants to shift the CBM to the lower values owing to the reduced bandgap energy value of ~2.34 eV. The appearance of secondary peaks for the N-doped TiO_2 model appeared due to the higher energy transitions from semi-core states in VB to CB.

The UV–Vis absorption spectrum is simulated to depict the optical properties of the N-doped TiO_2 anatase model. With the addition of the N-dopants, the electronic structure composition of the TiO_2 anatase model was changed which encompasses

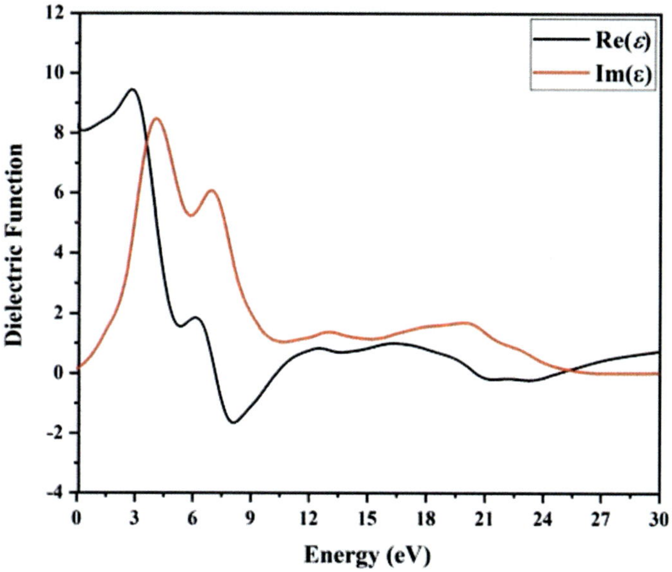

Fig. 4 The computed dielectric function presentation of the N-doped TiO_2 anatase model

the optical absorbance in the visible range. As reported in our previous work [23], The undoped anatase TiO_2 model exhibits a typical intense absorption peak in the UV region below 380 nm and is due to the intrinsic higher E_g of the TiO_2. It is notable to emphasize that the N-doped anatase TiO_2 model outspread the optical spectra to the visible range due to the insertion of the N-dopant impurities into the TiO_2 anatase crystal that reduces the bandgap and increased absorption of visible irradiation in solar spectra. The UV–Vis spectra display that the absorption edge extended up to a wavelength range of 438 nm in the visible region. For the N-doped TiO_2 anatase models, the energy gap transition of the electrons occurs from the hybridized O2p + N2p occupied states positioned in upper VB to the unoccupied Ti3d energy states in the lower CB. The results are also in accordance with our simulated electronic structure results. The N-doped TiO_2 anatase model thus also improves the absorption capacity of visible light photons, to ultimately produce more photogenerated e^-–h^+ pairs to signify the photocatalytic performance of the designed model. Accordingly, we can claim that the designed N-doped TiO_2 model is thus a practicable scheme to strengthen the photoactivity of the TiO_2 photocatalyst under visible irradiation (Fig. 5).

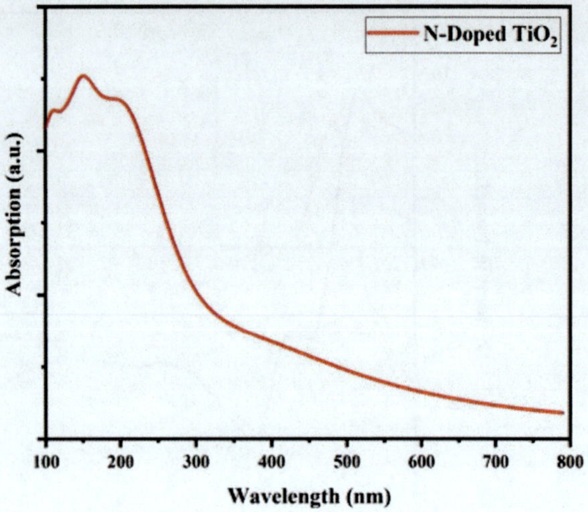

Fig. 5 The UV–Vis display of the designed N-doped TiO_2 anatase model

4 Conclusion

To conclude, the current work explored the effect of the N-doping concentration upon the structural and optoelectronic properties of the TiO_2 anatase photocatalyst in the frame of DFT calculations via GGA exchange correlational functional. For simulation of the structural properties, the GGA + PBE functional was implemented whereas to estimate the optoelectronic properties of the N-doped TiO_2 anatase models, the GGA + PBE + U functional agonized. The formation energies simulated for the designed models with Ti-rich conditions are lower than that associated with the O-rich conditions. Therefore, it indicates that the inclusion of N dopants into TiO_2 at the O atom site is more favorable than that at the Ti atom. The theoretical results conjecture that the N dopants impurity (induce energy) states in the electronic structure of the TiO_2 anatase model reduce the bandgap energy effectively and hence, improve visible light absorbance from the solar spectra. The bandgap was significantly reduced for the N-doped TiO_2 anatase model (~2.34 eV) as compared to 3.13 eV of the pure TiO_2 anatase. The absorption edge followed up to 438 nm for the N-doped TiO_2 anatase model. The enhanced visible light absorbance of the N-doped TiO_2 anatase model was associated with the comparatively more bandgap reduction for the N-doped system to have augmented photocatalytic activity. Therefore, the simulation results conclude that the N-doping technique of the TiO_2 anatase photocatalyst is more feasible and improves the photocatalytic performance efficiently.

Acknowledgements The authors acknowledge Grant # FRGS/1/2019/STG07/UTP/01/1, via Fundamental Research Grant Scheme (FRGS), and Universiti Teknologi PETRONAS (UTP) for financial assistance and lab facilities.

References

1. L. Weng, H. Zhang, A.O. Govorov, M. Ouyang, Hierarchical synthesis of non-centrosymmetric hybrid nanostructures and enabled plasmon-driven photocatalysis. Nat. Commun. **5**(1), 1–10 (2014)
2. M. Schiemann, B. Böhm, R. Chirone, O. Senneca, J. Ströhle, K. Umeki, M. Vujanovic, Technical solutions to foster the global energy transition: special issue on clean fuel conversion technologies for carbon dioxide and pollutant reduction, in *Book Technical Solutions to Foster the Global Energy Transition: Special Issue on Clean Fuel Conversion Technologies for Carbon Dioxide and Pollutant Reduction*, vol. 23 (2022), p. 111770
3. S. Reghunath, D. Pinheiro, S.D. KR, A review of hierarchical nanostructures of TiO_2: advances and applications. Appl. Surf. Sci. Adv. **3**, 100063 (2021)
4. J. Nowotny, T. Bak, M. Nowotny, L. Sheppard, Titanium dioxide for solar-hydrogen I. Functional properties. Int. J. Hydrog. Energy **32**(14), 2609–2629 (2007)
5. K. Batalović, N. Bundaleski, J. Radaković, N. Abazović, M. Mitrić, R.A. Silva, M. Savić, J. Belošević-Čavor, Z. Rakočević, C.M. Rangel, Modification of N-doped TiO_2 photocatalysts using noble metals (Pt, Pd)—a combined XPS and DFT study. Phys. Chem. Chem. Phys. **19**(10), 7062–7071 (2017)
6. A. Giannakas, M. Antonopoulou, C. Daikopoulos, Y. Deligiannakis, I. Konstantinou, Characterization and catalytic performance of B-doped, B-N co-doped and B–N–F tri-doped TiO_2 towards simultaneous Cr(VI) reduction and benzoic acid oxidation. Appl. Catal. B **184**, 44–54 (2016)
7. M. Islam, J. Podder, The role of Al and Co co-doping on the band gap tuning of TiO_2 thin films for applications in photovoltaic and optoelectronic devices. Mater. Sci. Semicond. Process. **121**, 105419 (2021)
8. H. Chen, X. Li, R. Wan, S. Kao-Walter, Y. Lei, C. Leng, A DFT study on modification mechanism of (N, S) interstitial co-doped rutile TiO_2. Chem. Phys. Lett. **695**, 8–18 (2018)
9. S.J. Clark, M.D. Segall, C.J. Pickard, P.J. Hasnip, M.I. Probert, K. Refson, M.C. Payne, First principles methods using CASTEP. Zeitschrift für Kristallographie-Crystal. Mater. **220**(5–6), 567–570 (2005)
10. J.D. Head, M.C. Zerner, A Broyden—Fletcher—Goldfarb—Shanno optimization procedure for molecular geometries. Chem. Phys. Lett. **122**(3), 264–270 (1985)
11. C. Xue, T. Zhang, D. Xiao, An advanced Broyden–Fletcher–Goldfarb–Shanno algorithm for prediction and output-related fault monitoring in case of outliers. J. Chem. **20**, 8–18 (2022)
12. N.M. Mohamed, F. Ullah, R. Bashiri, C.F. Kait, M.S.M. Saheed, M.U. Shahid, Hubbard's modified density functional theory calculations for the electronic structure and optical properties of carbon doped anatase TiO_2, in *Book Hubbard's Modified Density Functional Theory Calculations for the Electronic Structure and Optical Properties of Carbon Doped Anatase TiO_2*, vol. 45 (2021), pp. 371–381
13. E. German, R. Faccio, A.W. Mombru, Comparison of standard DFT and Hubbard-DFT methods in structural and electronic properties of TiO_2 polymorphs and H-titanate ultrathin sheets for DSSC application. Appl. Surf. Sci. **428**, 118–123 (2018)
14. F. Ullah, R. Bashiri, N. Muti Mohamed, A. Zaleska-Medynska, C.F. Kait, U. Ghani, M.U. Shahid, M.S.M. Saheed, Exploring graphene quantum dots@TiO_2 rutile (011) interface for visible-driven hydrogen production in photoelectrochemical cell: density functional theory and experimental study. Appl. Surf. Sci. **576**, 151871 (2022)
15. H.-C. Wu, S.-H. Li, S.-W. Lin, Effect of Fe concentration on Fe-doped anatase TiO_2 from GGA+ U calculations. Int. J. Photoenergy **25**, 391–401 (2012)
16. M. Berrahal, A. Bentouaf, H. Rached, R. Mebsout, B. Aissa, Investigation of Ruthenium based Full-Heusler compound for thermic, spintronics and thermoelectric applications: DFT computation. Mater. Sci. Semicond. Process. **134**, 106047 (2021)
17. R. Asahi, Y. Taga, W. Mannstadt, Electronic and optical properties of anatase. Phys. Rev. B Condens. Matter Mater. Phys. **61**(11), 7459–7465 (2000)

18. N.M. Mohamed, F. Ullah, R. Bashiri, C.F. Kait, M.S.M. Saheed, M.U. Shahid, Hubbard's modified density functional theory calculations for the electronic structure and optical properties of carbon doped anatase TiO_2, in *Book Hubbard's Modified Density Functional Theory Calculations for the Electronic Structure and Optical Properties of Carbon Doped Anatase TiO_2* (Springer, 2021), pp. 371–381
19. J. Safaei, H. Ullah, N.A. Mohamed, M.F.M. Noh, M.F. Soh, A.A. Tahir, N.A. Ludin, M.A. Ibrahim, W.N.R.W. Isahak, M.A.M. Teridi, Enhanced photoelectrochemical performance of Z-scheme g-C3N4/BiVO4 photocatalyst. Appl. Catal. B **234**, 296–310 (2018)
20. J. Wang, D.N. Tafen, J.P. Lewis, Z. Hong, A. Manivannan, M. Zhi, M. Li, N. Wu, Origin of photocatalytic activity of nitrogen-doped TiO_2 nanobelts. J. Am. Chem. Soc. **131**(34), 12290–12297 (2009)
21. R. Shirley, M. Kraft, O.R. Inderwildi, Electronic and optical properties of aluminium-doped anatase and rutile TiO_2 from ab initio calculations. Phys. Rev. B **81**(7), 075111 (2010)
22. H.H. Do, D.L.T. Nguyen, X.C. Nguyen, T.-H. Le, T.P. Nguyen, Q.T. Trinh, S.H. Ahn, D.-V.N. Vo, S.Y. Kim, Q. Van Le, Recent progress in TiO_2-based photocatalysts for hydrogen evolution reaction: a review. Arab. J. Chem. **13**(2), 3653–3671 (2020)
23. F. Ullah, N.M. Mohamed, M.S.M. Saheed, C.F. Kait, Density functional theory calculations for the electronic structure of carbon and copper co-doped TiO_2 rutile model, in *AIP Conference Proceedings*, vol. 2610(1) (2022), p. 030004

Natural Dye and Activated Carbon from Theobroma Cacao as Photosensitizer and Counter Electrode for Titania-Based Dye-Sensitized Solar Cell

Ronnel Delos Santos Magbitang, Siti Nur Azella Bt Zaine, Noridah Binti Osman, and Gerard Ang

Abstract Natural single-sourced dye-sensitized solar cells (DSSC) was successfully fabricated from Theobroma cacao. The activated carbon from cacao pod husks and the natural sensitizer derived from cacao leaves were utilized as a sustainable substitute for expensive platinum counter electrodes and N719 sensitizer to fabricate a cheaper and more eco-friendly DSSC. Cacao pod husks activated carbon (CPHAC) was prepared using a chemical activation process while the natural dye was extracted using the maceration technique. The morphology and textural properties of the synthesized CPHAC was characterized by different physiochemical techniques while optical characterization of the natural dye extracted from cacao leaves and synthetic N719 dye was conducted using UV visible spectrophotometry (UV–Vis). Decent photovoltaic parameters were recorded for the activated carbon counter electrode due to its significantly large surface area and highly porous morphological structures. The highest recorded efficiency was about 87.57% of the conventional DSSC.

Keywords Dye-sensitized solar cells · Activated carbon · Counter electrode · Chlorophyll sensitizer · Theobroma cacao

R. D. S. Magbitang (✉)
School of Graduate Studies, Mapua University, Intramuros Manila, Manila, Philippines
e-mail: rdmagbitang@mymail.mapua.edu.ph

S. N. A. B. Zaine
Centre of Innovative Nanostructures & Nanodevices (COINN), Universiti Teknologi PETRONAS, 32610 Bandar Seri Iskandar, Perak, Malaysia
e-mail: nurazella.zaine@utp.edu.my

N. B. Osman
Center for Biofuel and Biochemical Research (CBBR), Universiti Teknologi PETRONAS, 32610 Bandar Seri Iskandar, Perak, Malaysia
e-mail: noridah.osman@utp.edu.my

G. Ang
Electrical Engineering Department, School of Graduate Studies, Mapua University, Intramuros Manila, Manila, Philippines
e-mail: anggerard2k2@yahoo.com

M. B. Othman et al. (eds.), *Proceedings of the 1st International Conference of New Energy*, Springer Proceedings in Energy, https://doi.org/10.1007/978-981-99-0859-2_9

1 Introduction

Dye-sensitized solar cells (DSSC) are 3rd-era photovoltaic intended to be a promising device that straightforwardly converts solar energy into power. Generally, it comprises a dye-sensitized photoanode, an electrolyte with redox couple, and a counter electrode (CE). The counter electrode is responsible for speeding up the regeneration of iodide from triiodide and the regeneration of dye sensitizer molecules while sensitizers are responsible for light absorption which assimilate photons of a wavelength that corresponds to the energy distinction between its highest occupied molecular orbital (HOMO) and lowest unoccupied molecular orbital (LUMO) [1]. Platinum and ruthenium complex dye was the most commonly used synthetic materials for counter electrodes and sensitizers for DSSC, respectively. However, these materials are expensive, scarce, and not sustainable. Thus, investigating functional, stable, and superior Pt-free DSSC is vitally important [2]. Past and ongoing research has shown that the DSSC with activated carbon cathode terminal displayed a decent power conversion efficiency (PCE). The preparation of activated carbon from the wood of Choerospondias axillaris seed-stones and Alnus nepalensis plant was successfully implemented as activated carbon for DSSC which yield to 0.94% and 1.12% efficiency, respectively [3]. Moreover, natural dyes are viewed as the fundamental subbed, the least expensive, and sustainable sensitizers employed in the production of DSSC. These dyes can be extracted directly from plants that may contain chlorophyll, anthocyanin, carotenoids, tannins, cyanins, and flavonoids used as photosensitizers for DSSC [4]. A comparison study was conducted regarding the efficiency of chlorophyll dye extracted from pandan leaves using various solvents through the maceration technique. Results showed that chlorophyll in ethanol exhibited better efficiency compared to chlorophyll dye in acetonitrile, chloroform, ethyl ether, and methanol [5].

Theobroma Cacao, otherwise called the "food of the gods," cultivate bounteously in tropical nations like the Philippines. Cacao produces dried and aged greasy seeds, which are utilized to make chocolates [6]. More than half of the cacao's total weight was attributed to its pod and it was typically disposed of as waste or residue causing ecological problems like greenhouse gas emissions and foul scents. Nonetheless, cacao pod husks are wealthy in carbon sources proposing that they may be utilized as an energy source by ignition and changed over into valuable synthetic substances by pyrolysis [7]. Likewise, chlorophyll, being the most abundant pigmentation of plants can be extracted from its leaves. Thus, in this study theobroma cacao pod husk was used as a precursor for synthesizing activated carbon counter electrode while the natural dye extracted from theobroma cacao leaves was utilized as the photosensitizer for titania-based DSSC since there is no existing study regarding the utilization of Theobroma cacao for single-sourced DSSC. The implementation of this natural dye and activated carbon cathode is propitious due to its financial and environmental characteristics since it is single-sourced [8]. Furthermore, the photovoltaic characteristics of activated carbon-based DSSC, chlorophyll-based DSSC, and cacao-based DSSC were compared to the conventional DSSC.

2 Methodology

2.1 Synthesis of Cacao Pod Husk Activated Carbon (CPHAC)

The cacao pod husk wastes were collected from the Lembaga Koko Malaysia at Bagan Datuk, Perak, Malaysia. The pre-washed fresh cocoa pod husks were sun-dried for 24 h and crushed before further drying in the oven at 80 °C until a constant weight was obtained (Memmert, UN55). The dried cacao pod husks were grounded (Retsch, SM100 Rostfrei) and subsequently sieved to fractions with particle size 0.5 mm (Endecott's Ltd, Minor-0646–03). The CPH powder was carbonized using a drop-type fixed-bed reactor at 500 °C in an inert nitrogen atmosphere and impregnated using potassium hydroxide (KOH) utilizing 0.5 g KOH/g CPH impregnation ratio at 30 °C for 1 h before drying in an oven at 110 °C for 24 h. Subsequently, the dried impregnated sample was further pyrolyzed at 800 °C using a tube furnace with the presence of nitrogen gas. Then, the KOH-treated sample was neutralized using 3 M hydrochloric acid (HCl) and deionized water. Finally, the resulting product was dried overnight in an oven at about 110 °C and stored in an air-tight container until further use [7] (Fig. 1).

Fig. 1 Activated carbon production from cacao pod husks

Fig. 2 Natural dye extraction from cacao leaves via maceration technique

2.2 *Sensitizers Preparation*

Fresh cacao leaves were obtained from Lembaga Koko Malaysia at Bagan Datuk, Perak. About 200 g of the pre-washed cacao leaves were combined with absolute ethanol utilizing 1:10 g/ml solid-to-liquid ratio. The samples were macerated for a week in a dark room before filtration (Whatman no. 1, 150 mm diameter). Solvent evaporation was done using a rotary evaporator at 70 °C until completely dried and then 200 ml of absolute ethanol was added to collect the extracted dye. Finally, the dye solution was properly stored in a vial and placed in the refrigerator to protect it from direct sunlight and atmospheric air. For the synthetic dye solution, about 0.755 g of N719 powder was dissolved in 2.5L of absolute ethanol to make 0.254 mM N719 dye solution (Fig. 2).

2.3 *Fabrication of Dye-Sensitized Solar Cell*

2.3.1 Preparation of Working Electrode

The pre-washed Fluorine-Tin-Oxide (FTO) glass was treated with 40 mM $TiCl_4$ aqueous solution at 70 °C for 30 min and washed with deionized water and ethanol. The TiO_2 with a thickness of 6 μm was applied twice on the FTO glass to make an active area of 1 cm^2 using screen printing method (ATMA, AT-60PD). The TiO_2 was dried before printing with another layer using 4 zones of conveyor furnace (HENGLI, HSH3003-0402, approx. 65" of belt working length) with a speed of 124 mm/min. The drying temperature was set at 75 °C, 85 °C, 95 °C and 85 °C for the first, second, third, and fourth zone, respectively. After applying the second layer, the working electrodes were cut into small cells and fired in a 6 zones conveyor furnace

(HENGLI, HSH3003-0407, approx. 119.5" of belt working length) at temperatures of 300 °C, 510 °C, 505 °C, 350 °C, 270 °C and 180 °C for a consecutive zone with a conveyor speed of 140 mm/min. The working electrodes were treated again with $TiCl_4$ aqueous solution and refired in a 6 zones conveyor furnace as described above. The TiO_2-coated glass substrates were removed from the furnace belt and immersed in the staining solution of natural dye and synthetic dye, respectively. Finally, the electrodes were removed from the staining bath and carefully rinsed with ethanol.

2.3.2 Preparation of Counter Electrode

Platinum Counter Electrode

The platinum paste was screen printed onto the pre-washed and drilled FTO counter electrode. The Pt layer was dried with the same parameter as the working electrodes sides, then heat treated using a 4 zones conveyor furnace (HENGLI, HGL3005-4A, approx. 457" of belt working length) at speed of 148 mm/min and temperature of 350 °C, 410 °C, 240 °C and 190 °C for each zone, respectively.

Activated Carbon Counter Electrode (ACCE)

The synthesized CPHAC was dried in a vacuum oven at 70 °C overnight to minimize its moisture content. 10 mg of activated carbon, carbon black, and polyvinylidene fluoride were combined using an 80:10:10 ratio. The mixture was grounded for 20 min using agate mortar and pestle. To make the activated carbon slurry, 300 μl of 1-methyl-2-pyrrolidinone (NMP) was added to the mixture and grounded again for 10 min. The slurry was then deposited on pre-washed FTO glass using the D-blade technique. Finally, the counter electrode was sintered at 80 °C overnight using the oven.

2.4 Dye-Sensitized Solar Cell Assembly

The working electrode and counter electrode were combined into a sealed sandwich-type cells using a thermoplastic gasket via a compression heat process. Two drops of redox iodide/triodide electrolyte (HPE, Dyesol) were inserted through sandblasted holes on the counter electrodes via vacuum backfilling. The holes are then closed by aluminum backed bynel. Finally, cerazoler active solder alloy was applied as a current collector using an ultrasonic soldering system.

Table 1 List of fabricated dye-sensitized solar cells

DSSC	Sensitizer(s)	CE	Electrolyte
Conventional	N719	Platinum	HPE
AC-based	N719	ACCE-CB	HPE
Chlorophyll-based	Chlorophyll	Platinum	HPE
Cacao-based	Chlorophyll	ACCE-CB	HPE

2.5 Characterization of the Dye-Sensitized Solar Cells

Field Emission Scanning Electron Microscope and Energy-dispersive X-ray Spectroscopy (FESEM-EDx) was conducted to investigate the surface morphology and elemental composition of the synthesized CPHAC (Tescan, Clara and Oxford Instrument, Ultim Max). The pore properties and surface area of the CPHAC were analyzed using surface area and porosity analysis (Micromeritics, Tristar II 3020). The optical characterization of the dye was conducted using UV–Visible spectrophotometry (Cary 60) to analyze the light absorbance of N719 and natural dye from cacao leaves. The performance of the fabricated DSSC and silicon-based PV cell was evaluated using a Universal Photovoltaic Test System under 100 $mWcm^{-2}$ intensity of an illuminant Xenon lamp at an AM 1.5 radiation angle connected to a voltmeter and ampere meter (Model 2420, Keithly).

3 Results and Discussions

3.1 Field Emission Scanning Electron Microscopy

Figure 3 shows FESEM micrographs of synthesized CPHAC. Images revealed the developed porous structure and the different morphologies of the synthesized activated carbon. Images a.1 and a.2 exhibit the spongy structure of the activated carbon. The development of macropores which are approximately 3–7 microns (measured using FESEM image software) are shown in images b.1 and b.2. It has been noticed that it consists of shallow pores that can be related to the carbonization of the CPHAC at 500 °C which resulted in the creation of initial porosity on the surface of the material [9, 10]. Images c.1 and c.2 manifest the formation of mesopores and micropores of CPHAC. The development of mesopores and micropores can be attributed to further heating of the CPH at 800 °C [11, 10, 12–16]. Images d.1 and d.2 show that the material consists of numerous deeper and tinier holes which are roughly 1 micron (measured using FESEM image software) which may be due to KOH treatment since chemical impregnation allows deep penetration into carbon structure [17, 18].

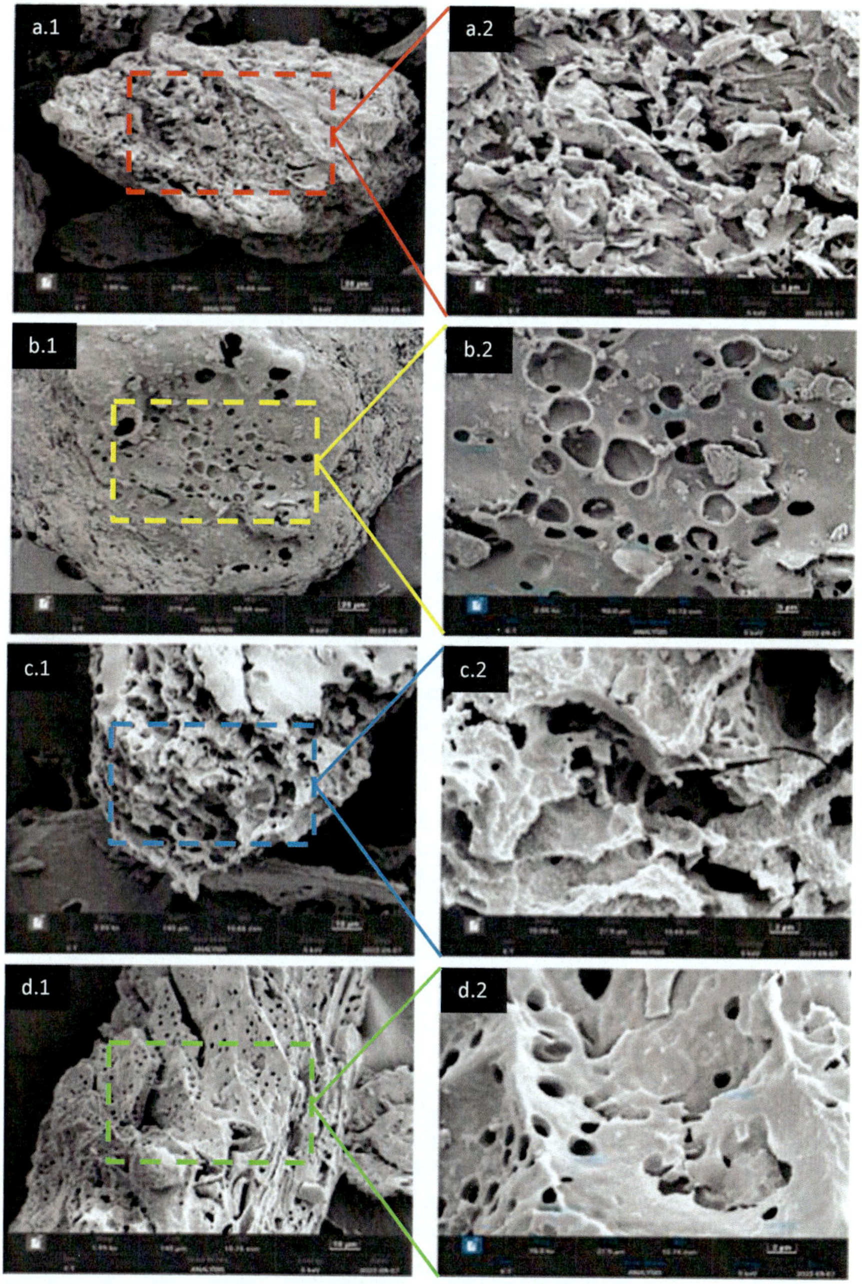

Fig. 3 Various surface morphology of CPHAC obtained through FESEM

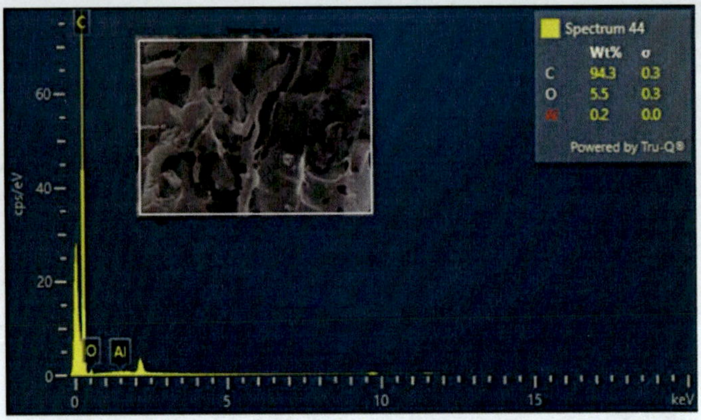

Fig. 4 EDX spectrum of CPHAC

3.2 *Energy-Dispersive X-ray Spectroscopy (EDX)*

Figure 4 shows the EDX spectrum of the synthesized CPHAC that contains 94.3 wt% carbon, 5.5 wt% of oxygen, and 0.02 wt% of Aluminum. This indicates that approximately one in every 19 carbon atoms has oxygen features [19]. The presence of aluminum can be related to the sample holder used for the analysis and/or the equipment used in the pyrolysis process while the presence of oxygen is due to the partial decomposition of the oxygen content of the precursor during pyrolysis or oxygen-bonding occurs while activation process [20, 21]. Thus, EDX reveals that the synthesized CPHAC has high carbon content and the absence of impurities that may affect the performance of the material as a counter electrode for the DSSC such as potassium and chlorine used for impregnation pretreatment of the CPH and neutralizing the activated carbon, respectively.

3.3 *Surface Area and Porosity Analysis*

The nitrogen adsorption–desorption graph of the synthesized CPHAC as illustrated in Fig. 5a can be characterized as a combination of Type I and Type IV isotherms based on the International Union of Pure and Applied Chemistry (IUPAC) classification indicated by small H4 hysteresis loop at $P/Po > 0.44$. Type I isotherm referred to microporous material while Type IV isotherm corresponds to mesoporous materials [22, 23]. The large quantity of microporous feature of the sample can be confirmed by adsorption indicated by a significant increment in adsorption at very low relative pressure and the presence of mesopores was confirmed by the slope from 0.44–1.0 P/Po [24] shown in Fig. 5a. Detailed adsorption isotherm in micropore region was demonstrated in Fig. 5b. Brunauer–Emmett–Teller (BET) method by multi-point technique was used to estimate the specific surface area which was found to be

946.7420 m^2/g through mesopore analysis and 960.4312 m^2/g using micropore analysis are shown in Fig. 5c, d. Additionally, Barrett–Joyner–Halenda (BJH) method was used to determine the pore size distribution of the sample. Figure 5e shows the pore distribution plot in the mesoporous region. The total pore volume, average pore diameter, and pore of the sample were 0.502420 cm^3/g, 4.0475 nm, and 2.12273 nm, respectively. Pore development that ranges from 0.64–1 nm with a peak at 0.68 nm was observed and analyzed from the pore distribution in the microporous region using Original Density Functional Theory (DFT) as shown in Fig. 5f [25]. Hence, SAP analysis confirms that the synthesized CPHAC has a large surface area and is highly porous. Such characteristics are beneficial properties in producing efficient carbon-based dye-sensitized solar cells such as high surface area, good electrocatalytic activity, and conductivity which may contribute to the efficient charge transfer, electron collection, and ion diffusion [26–28].

3.4 UV–Visible Spectroscopy

It is essential to study the absorption capability of the sensitizer for the general performance of the DSSC can be correlated to the light absorption capability of the dye for electron excitations and conversion of light to power. Based on UV visible spectrum of the two distinct dyes as shown in Fig. 6, four absorbance peaks were observed for the synthetic N719 dye, the first two peaks located in the ultraviolet region specifically at 235 nm and 265 nm corresponding to the π-π^* charge transfer transition while the other two peaks in the visible region spotted at 388 nm and 531 nm were attributed to metal-to-ligand charge transfer (MLCT) origin [29–31]. Moreover, two peaks were recorded for cacao leaves dye. One in the near violet-blue visible region (352 nm) can be attributed to π-π^* transition of the conjugated bonds and intermolecular charge transfer (ICT) transitions and another peak at the orange-red visible region (637 nm) which could be due to the n- π^* electron transition. Furthermore, the absorption spectra of the extracted dye from cacao leaves indicate the presence of chlorophyll [32–34].

3.5 Photocurrent–voltage Measurement

The photovoltaic characteristics of the fabricated DSSC with a 1 cm^2 active area were measured in 100mWcm^{-2} illumination. Parameters such as efficiency (Eff), fill factor (ff), open circuit voltage (Voc), short circuit current (Jsc), maximum voltage (Vmax), and maximum current (Jmax) were tabulated in Table 2. All photocurrent–voltage measurements were conducted right after the fabrication of the device (Fig. 7).

Activated carbon counter electrodes exhibited very promising photovoltaic performance as shown in Table 2. The efficiency of activated carbon-based DSSC

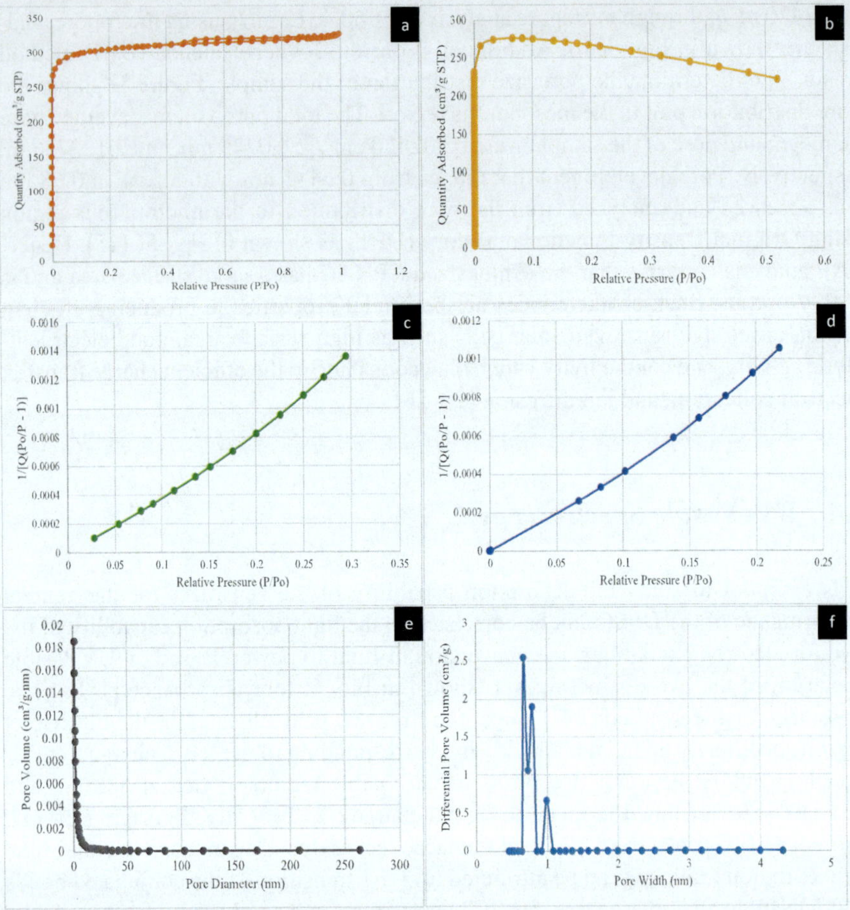

Fig. 5 (**a**) Adsorption–Desorption Isoterms through mesopore analysis (**b**) Adsorption Isoterms through micropore analysis (**c**) BET surface plot through mesopore analysis (**d**) BET surface plot through micropore analysis (**e**) Mesopore distribution plot using BHJ method (**f**) Micropore distribution curve using DFT method

(AC-based DSSC) was about 87.57% of the conventional DSSC. The high efficiency of AC-based DSSC is primarily due to the substantial surface area, highly porous morphology, and oxygen-containing surface functional group which can be confirmed by surface area and porosity analysis and EDX results, respectively. Moreover, the fill factor, open circuit voltage, and maximum voltage of the activated carbon-based DSSC were found to be larger compared to the conventional DSSC [24, 35]. Greater Voc values of activated carbon-based DSSC than Pt-based DSSC can be related to the lower electron recombination rate or to the higher number of electrons being inserted into the TiO_2 conduction band since Voc relies normally on this occurrence [36–38]. However, the short circuit and maximum current of

Fig. 6 UV–Vis Spectrum of N719 and Chlorophyll

Table 2 Comparison of photovoltaic parameters of conventional, activated carbon-based, chlorophyll-based, and cacao-based DSSC under $100mWcm^{-2}$

DSSC	EFF (%)	FF (%)	Voc (mV)	Jsc ($mAcm^{-2}$)	Vmax (mV)	Jmax ($mAcm^{-2}$)
Conventional	5.55	39.70	563.60	23.53	300.50	17.54
AC-based	4.86	39.80	611.80	18.96	320.00	14.42
Chlorophyll-based	0.34	70.10	441.80	1.05	339.00	0.96
Cacao-based	0.08	25.20	411.70	0.74	165.80	0.46

conventional DSSC were better than the activated carbon-based DSSC because metal-based electrodes such as platinum have better conductivity [24, 35]. Moreover, chlorophyll-based DSSC only produced 0.34% efficiency. The low efficiency of the device utilizing chlorophyll dye can be related to the weak bond between the carboxylic functional groups of chlorophyll dye molecule and the photoanode which is responsible for efficient charge injection. Such a strong bond between dye and photoanode prevents charge outflow in electrolytes [39–41]. Additionally, the device photovoltaic performance was comparable to a previous study conducted utilizing chlorophyll dye. Furthermore, cacao-based DSSC only produced 0.08% efficiency.

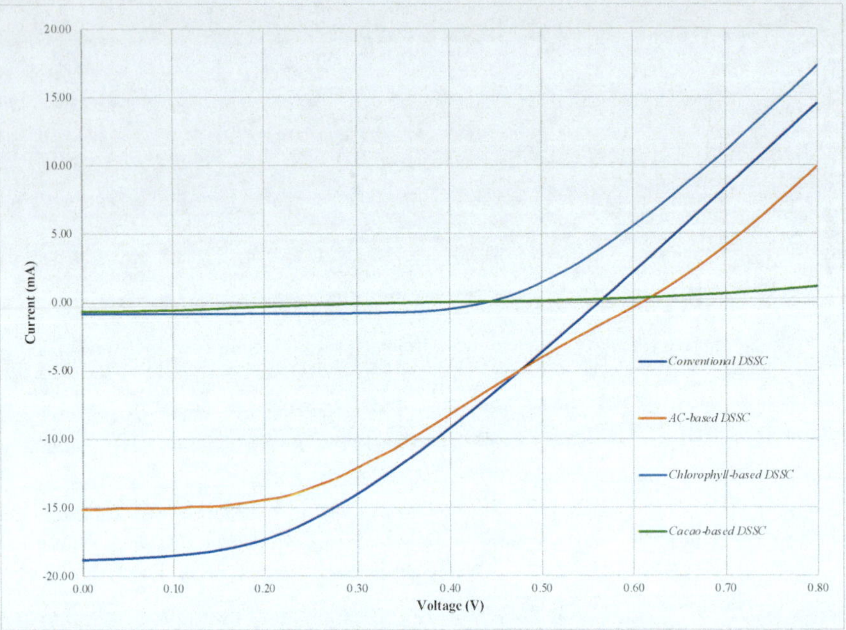

Fig. 7 IV Curves under 1 sun

4 Conclusion

In this study, activated carbon from Theobroma cacao pod husks and chlorophyll derived from Theobroma cacao leaves were utilized as a sustainable substitute for expensive platinum counter electrodes and N719 sensitizer to fabricate a cheaper and more eco-friendly DSSC. It was found that AC-based DSSC utilizing the synthesized activated carbon from Theobroma cacao pod husk recorded a decent light-to-power transformation efficiency which is very comparable to Pt-based DSSC. However, the natural dye from cacao leaves only produced 0.34% efficiency. Moreover, a very low efficiency was recorded for cacao-based DSSC. Thus, activated carbon can be a potential substitute for platinum but chlorophyll cannot be used to replace the synthetic N719 dye. Furthermore, the future study may focus on the effectivity of chlorophyll as a cosensitizer for N719 under lower light conditions and the effect of activated carbon counter electrode preparation on the DSSC photovoltaic performance.

Acknowledgements This is to acknowledge the Department of Science and Technology-Engineering Research and Development for Technology (DOST-ERDT) for funding this research and Yayasan Universiti Teknologi PETRONAS through Fundamental Research Grant (015LC0-449) for supporting the submission of this study on 1st International Conference on New Energy (ICNE2022) under World Engineering, Science & Technology Congress (ESTCON 2022).

References

1. N. Y. Amogne, D. W. Ayele and Y. A. Tsigie, Recent advances in anthocyanin dyes extracted from plants for dye sensitized solar cell, Mater. Renew. Sustain. Energy, p. 9:23, (2020)
2. Jingwen Li et al, Biomass-derived carbon boosted catalytic properties of tungsten-based nanohybrids for accelerating the triiodide reduction in dye-sensitized solar cells, J. Colloid Interface Sci. (2020)
3. P. Joshi, Novel counter electrodes of dye-sensitized solar cells based on activated carbon prepared from wood of Choerospondias axillaris seed-stones and Alnus nepalensis plant, *International Journal of Engineering and Advanced Research Technology (IJEART)*, **3**(3), pp. ISSN: 2454–9290, (March 2017)
4. Zhou, Wu, Gao and Ma, Dye-sensitized solar cells using 20 natural dyes as sensitizers., J. Photochem. Photobiol. A Chem. **219**(2), p. 188–194, (2011)
5. M. A. M. Al-Alwani et. al, Natural dye extracted from Pandannus amaryllifolius leaves as sensitizer in fabrication of dye-sensitized solar cells, Int. J. Electrochem. Sci., pp. 747–761, (2017)
6. S. Palisoc et. al, Remediation of water using *Theobroma Cacao* pod-based activated carbon: A novel approach as determined by highly sensitive BiNP/MWCNT/Nafion-modified GCE, Mater. Res. Express, (2019)
7. W.-T. Tsai et. al, Porous and adsorption properties of activated carbon prepared from cocoa pod husk by chemical activation, Biomass Convers. Biorefinery, (2019)
8. K. Gasga, Dye and activated carbon from Canarium ovatum Engl. as photosensitizer and counter electrode for Titania-based dye sensitized solar cell, J. Phys.: Conf. Ser. 1882 012007, (2021)
9. A. W. S. M. Daud WMAW, The effects of carbonization temperature on pore development in palm-shell-based activated carbon. Carbon, **38**, p. 1925–32, (2000)
10. M.-S. M. Rodriguez-Reinoso F, Activated carbons from lignocellulosic meterials by chemical and/or physical activation: an overview. Carbon, **30**(7), p. 1111–8, (1992)
11. Z. A.-Q. C. Z. N. Mohd Adib Yahya, Agricultural bio-waste materials as potential sustainable precursors used for activated carbon production: A review, Renew. Sustain. Energy Rev. **46**, p. 218–235, (2015)
12. K. A. Z. Z. H. M. K. D. A. F. e. a. Abdullah AH, Preparation and characterization of activated carbon from Gelam Wood Bark (Melaleuca cajuputi), Malays J. Anal. Sci. vol. **7**(1), p. 65–8, (2001)
13. M.-S. M. R.-R. F. Caturla F, Preparation of activated carbon by chemical activation with ZnCl2. Carbon, **29**(7), pp. 999–1007, (1991)
14. S. F. D. C. L. R. V. A. S. K. Solar C, Natural gas storage in microporous carbon obtained from waste of the olive oil production, Mater. Res. **11**(4), pp. 409–14, (2008)
15. P. P. S. P. Baseri JR, Preparation and characterization of activated carbon from Thevetia peruviana for the removal of dyes from textile waste water, Adv. Appl. Sci. Res. **3**(1), pp. 377–83, (2012)
16. Y. K. P. J. Z. L. G. S. X. H. Li W, Effects of carbonization temperatures on characteristics of porosity in coconut shell chars and activated carbons derived from carbonized coconut shell chars, Ind. Crops Prod. **28**, pp. 190–8, (2008)
17. M. H. D. M. H. G. J. I. A. M. S. Zoha Heidarinejad, Methods for preparation and activation of activated carbon: a review, Environ. Chem. Lett. pp. https://doi.org/10.1007/s10311-019-00955-0, (2019)
18. P. T. C. R.-A. e. a. Gratuito MKB, Production of activated carbon from coconut shell: optimization using response surface methodology, Bioresour Technol., **99**, p. 4887–4895 https://doi.org/10.1016/j.biortech.2007.09.042, (2008)
19. G. K. R. R. D. L. K. T. K.D.M.S.P.K. Kumarasinghe, Activated coconut shell charcoal based counter electrode for dye-sensitized T solar cells, Org. Electron., **71**, pp. 93–97, (2019)
20. A. Afrianda, Pembuatan dan Karakterisasi Elektroda Karbon Sel superkapasitor dari Bahan Ampas Sagu Menggunakan Aktivasi H2O Berdasarkan Variasi Waktu Aktivasi, Skripsi, Jurusan fisika fakultas matematika dan ilmu pengetahuan alam universitas riau, p. 201, (2018)

21. S. E. R. a. R. Y. Y Yuli, Biomass waste of cocoa skin for basic activated carbon as source of eco-friendly energy storage, J. Phys.: Conf. Ser., pp. doi:https://doi.org/10.1088/1742-6596/1788/1/012020, (2021)
22. A. V. I. L. P. G. L. D. L. A. P. T. R. C. J. Sabrina F. Lütke, Preparation of activated carbon from black wattle bark waste and its application for phenol adsorption, J. Environ. Chem. Eng., **7**, p. https://doi.org/10.1016/j.jece.2019.103396, (2019)
23. G. Singh et al., Convenient design of porous and heteroatom self-doped carbons for CO2 capture. Microporous Mesoporous Mater. **287**, 1–8 (2019)
24. G. N. S. C. S. L. K. B. J. S. Y. Sung Min Cha 1, Fallen leaves derived honeycomb-like porous carbon as a metal-free and low-cost counter electrode for dye-sensitized solar cells with excellent tri-iodide reduction, J. Colloid Interface Sci., **513**, pp. 843–851, (2018)
25. N. S. N. M. A. A. Z. U. D. H. M. Z. F. N. A. Jibril Mohammed, Optimization of microwave irradiated—coconut shell activated carbon usingresponse surface methodology for adsorption of benzene and toluene, Desalination Water Treat., pp. 7881–7897, (2015)
26. H. A. Z. A. M. Jaafar, The use of carbon black-TiO2 composite prepared using solid state method as counter electrode and E. conferta as sensitizer for dye-sensitized solar cell (DSSC) applications, Opt. Mater. **79**, pp. 366–371, (January 2018)
27. H. A. M. A. Z. Jaafar, Performance of dye-sensitized solar cell (DSSC) using carbon black-TiO2 composite as counter electrode subjected to different annealing temperatures, Opt. Quant. Electron., **52**(4), (2020)
28. P. E. Y. W. S. H. N. C. M. L. Nurnajaa Narudin, Enhanced properties of low-cost carbon black-graphite counter electrode in DSSC by incorporating binders, Sol. Energy, **225**, pp. 237–244, (2021)
29. M. K. Nazeeruddin, E. Baranoff and M. Gra¨tzel, Dye-sensitized solar cells: A brief overview, Sol. Energy, pp. 1172–1178, (2011)
30. S. Agarwala, M. Kevin, A. Wong, C. Peh, V. Thavasi, G. Ho, Mesophase ordering of TiO2 film with high surface area and strong light harvesting for dye-sensitized solar cell. ACS Appl. Mater. & Interfaces **2**, 1844–1850 (2010)
31. M. U. Shahid, S. N. Azella Zaine, N. Muti Mohamed, M. Khatani, A. E. Samsudin, Trap state and charge recombination in nanocrystalline passivized conductive and photoelectrode interface of dye-sensitized solar cell, Coatings, **10**, p. 284, (2020)
32. K.H.M. Younas, Performance enhancement of dye-sensitized solar cells via co-sensitization of ruthenium (II) based N749 dye and organic sensitizer RK1. Sol. Energy **203**, 260–266 (2020)
33. *. C. N. H. a. M. A. S. b. J. F. W. c. A. N. O. d. N. N. A. e. Sabastine C. Ezike a, S. C. Ezike, C. N. Hyelnasinyi, M. A. Salawu, A. N. Ossai, J. F. Wansah and N. N. Agu, Synergestic effect of chlorophyll and anthocyanin Co-sensitizers in TiO2-based dye-sensitized solar cells, Surfaces and Interfaces, (22), p. 100882, (2021)
34. R. Nakhaei, A. Razeghizadeh, P. Shabani, J. Ganji and S. S. Tabatabaee, Photoabsorption enhancement in synthetic-natural dye- sensitized solar cells using bilayer TiO2 deposition and separated sensitization, Hindawi: Int. J. Photoenergy, p. https://doi.org/10.1155/2022/5949837, (2022)
35. J. H. L. S. M. C. J. S. Y. Goli Nagaraju, Three-dimensional activated porous carbon with meso/macropore structures derived from fallen pine cone flowers: A low-cost counter electrode material in dye-sensitized solar cells, J. Alloy. Compd., **693**, pp. 1297–1304, (2017)
36. V. S. a. P. B. Rahul Kumar, Fabrication of a counter electrode for dye- sensitized solar cells (DSSCs) using a carbon material produced with the organic ligand 2- methyl-8-hydroxyquinolinol (Mq), Nanoscale Adv., **1** pp. 3192, (2019)
37. M. T. Z. P. K. T. M.-M. u. R. H. B. J. Butt, Biomass-derived nitrogen-doped carbon aerogel counter electrodes for dye sensitized solar cells, *Materials,* **11**, p. 1171, (2018)
38. A.M.K.S.N.C.V.S.S.N.B.Y.V.S.K.V.M.R.M. Gurulakshmi, A transparent and Pt-free all-carbon nanocomposite counter electrode T catalyst for efficient dye sensitized solar cells. Sol. Energy **193**, 568–575 (2019)
39. G. M. Kay A, Artificial photosynthesis. 1. Photosensitization of titania solar cells with chlorophyll derivatives and related natural porphyrins, J. Phys. Chem., **97**(23), p. 6272–7. https://doi.org/10.1021/j100125a029, (1993)

40. C. I. D. M. G. A. M. S. M. M. G. G. Calogero G, Brown seaweed pigment as a dye source for photoelectrochemical solar cells., Spectrochim Acta – Part A: Mol Biomol Spectrosc , **117**, pp. 702–6, (2014)
41. S.M.M.M.B.S.A.H.G.A.H.M.F.G.T.D.S.M.R.M.R.A.A.A.A.M.A.M.M.A.F. Kabira, Instability of dye-sensitized solar cells using natural dyes and approaches to improving stability—An overview. Sustainable Energy Technol. Assess. **52**, 102196 (2022)

Solvation-Free Energy and Thermodynamic Properties of Hydrogen Adsorption Inside Porous HKUST-1 Composite Through Molecular Dynamics Simulation

Mohamad Adil Iman Bin Ishak, Nor Ain Fathihah Binti Abdullah, Khairulazhar Bin Jumbri, and Mohd Faisal Bin Taha

Abstract Hydrogen as one of the renewable energy resources has unlocked the future potential for the economical and eco-friendly energy supply to meet the world demand in various sectors. Principal on hydrogen adsorption and storage can be complex depending on the material selection, thus, it is crucial to fully understand its molecular mechanics for the sustainable energy transition. This study is conducted to fulfill the missing information on the hydrogen adsorption inside nano-porous HKUST-1 composite through solvation-free energy calculation and highlighting their thermodynamics properties to explain the details on the molecular interaction between H_2 and HKUST-1 where all the data and calculations are computed using GROMACS package. At the end of this study, molecular information on the H_2@HKUST-1 composite will be fully discussed and that information can be used as the basic guideline for future improvement.

Keywords Hydrogen adsorption · Metal–organic frameworks · Molecular dynamics simulation

1 Introduction

Technological advancement and economic growth have increased energy consumption around the world, coupled with the increase in population, industrial development, and the improvement of social standards in society. This causes the need for energy to become higher to accommodate the demand from all networks and sectors. However, the energy base that is used in the past, which is a fossil fuel, has been decreasing and shrinking from year to year. Therefore, new alternative energy is needed to replace existing energy sources. Looking at the sustainable aspects of the environment, hydrogen energy is one of the potential alternatives to consider. What

M. A. I. B. Ishak · N. A. F. B. Abdullah · K. B. Jumbri · M. F. B. Taha (✉)
Department of Fundamental and Applied Sciences, Universiti Teknologi PETRONAS, 32610 Bandar Seri Iskandar, Perak, Malaysia
e-mail: faisalt@utp.edu.my

M. B. Othman et al. (eds.), *Proceedings of the 1st International Conference of New Energy*, Springer Proceedings in Energy, https://doi.org/10.1007/978-981-99-0859-2_10

is interesting about this energy is that it is safe, renewable, obtained through various chemical reactions, and has a high energy capacity. However, the understanding and exploration of this energy is still low and lacks sufficient of expertise. The intended understanding is the mechanism of hydrogen absorption, transportation, storage, and reuse.

Many mediums and materials have been proposed to be used as hydrogen adsorbers, but not all have shown good and satisfactory results. Metal–organic frameworks (MOFs) are one of the most talked about adsorbent materials because it has interesting physical properties compared to other absorbents. Furthermore, it has been proven to be a good material for a wide range of applications such as catalysts [1], chemical sensors [2], batteries and electrolytes [3], gas adsorbent [4], gas storage [5], and drug delivery [6]. Their role in gas adsorption is continuously explored in various studies. Parkes et al. [7] conducted a computational study on the effect of MOF flexibility toward gas diffusion in HKUST-1 and ZIF-8. They found that gas diffusion was higher in the flexible framework compared to the rigid framework. Plus, the diffusion of gases in HKUST-1 was affected by two competing effects. The first one was the steric effect where the diffusion rate decreased with the increase of gas loading meanwhile the second effect called the 'small caged effect' showed increasing in gas diffusion along with the gas loading. Recent study done by Han et al. [5] showed thermodynamic properties of gases inside MOFs for energy conversion. Final outcome showed that enthalpy change was affected by the pore width while entropy change of gases was influenced by both pore width and volume. Both data showing maximum value at 4.0 and 7.5 Å of pore size. They proposed that thermodynamic properties of gases would provide a useful guideline for choosing suitable adsorbent-adsorbate for heat transformation and gas storage.

Based on this principle, this study was conducted by investigating the dynamic properties of H_2 adsorption inside HKUST-1 framework through molecular dynamic simulations. HKUST-1 is chosen in this study as it was proven to capture a large amount of hydrogen and have high thermal strength. Chui et al. [8] initially announced the existence of HKUST-1 in 1999. It is made up of 1,3,5-benzenetricarboxylate (BTC) ligands that coordinate copper ions in a cubic lattice. At 77 K, it has been demonstrated that HKUST-1 can adsorb large amounts of hydrogen. However, even at room temperature or above, the adsorption capacity is not fully revealed which this information is crucial for certain applications, especially for mobile and transportation. Therefore, the primary goals of this work were to determine the capability of HKUST-1 as a hydrogen storage material at higher temperatures.

2 Methodology

The open-access online database Chem Tube 3D was used to determine the crystal structure of HKUST-1 [9]. As seen in Fig. 1, one subunit, HKUST-1, is basically made up of two node copper atoms functioning as the metal center bonded to benzene-1,3,5-tricarboxylate (BTC) linker molecules. Ten hydrogen molecules were positioned in

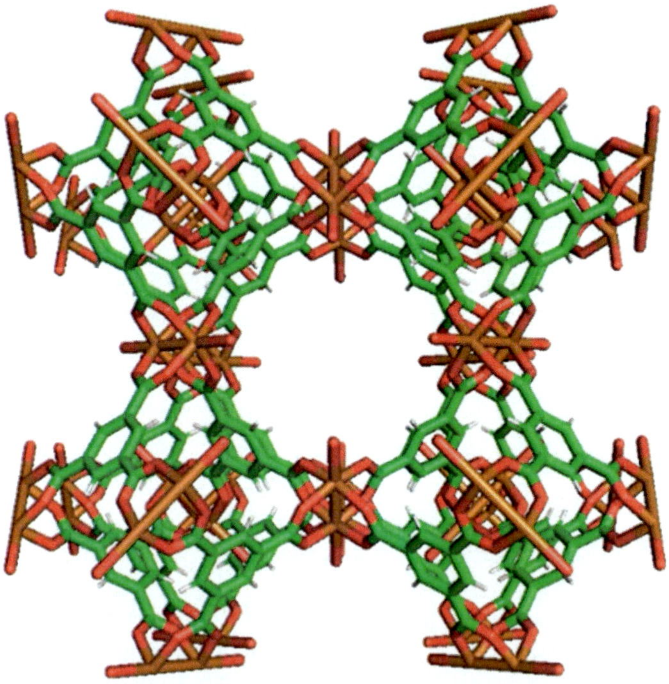

Fig. 1 Crystallographic structure of HKUST-1 used in the simulation. Color notation of atom: Copper (brown), Oxygen (red), Carbon (green), Hydrogen (white)

the center of the HKUST-1 unit cell, which was built in cubic dimensions with a box size of 2.6 nm × 2.6 nm × 2.6 nm. The GROMACS package's Packmol [10] component was used to pack HKUST-1 and hydrogen molecules together. Since just one subunit, HKUST-1, was employed in the simulation, the number of hydrogen molecules to be packed was fixed at ten, adopting from the previous study by Ishak et al. [11]. This was also done to ensure proper loading.

The OBGMX [12] topology generator produced the entire hydrogen molecule and HKUST-1 topology. Adapting from Chen et al. [13], the forcefield was used to alter the produced topology. The surplus chemical potential was then determined using the Solvation Free Energy Bennet Acceptance Ratio (BAR) method. Twenty-one sets of parameters, with values ranging from $\lambda = 0$ to $\lambda = 1$, were used, with a five nanosecond total simulation period for each system. The long-range electrostatic interactions were addressed using Particle-Mesh Ewald (PME) [14] with a grid spacing of 0.12 nm and fourth-order interpolation, whereas the nonbonded interactions were estimated up to 1.2 nm. Motion integration took place over a step of 2.0 fs. The neighbor search, which was updated every five steps, had a range of up to 1.2 nm. The bond lengths were restricted using LINCS [15]. Temperature and pressure were controlled, respectively, by the Berendsen thermostat and barostat [16]. We used a reference pressure of 1.0 atm and a relaxation time of 2.0 ps. The

Table 1 The density value of HKUST-1 from simulated and experimental data at 298 K and 1.0 bar

No.	Parameter	Simulated (this work)	Experimental	% error
1	Density (g/cm^3)	0.88	0.88 [17]	0.00

isothermal compressibility of the pressure control was tuned to 4.5×10^{-5} bar^{-1}. The heat was separated using two heat baths with temperature coupling constants of 0.1 ps. The systems began by minimizing energy in 5000 steps for each steepest descent and conjugate gradient. A conical ensemble (NVT) was then used for a further 100 ps to bring the system to equilibrium at 298 K and 1.0 bar, followed by an isothermal-isobaric system (NPT) for 5 ns.

3 Result and Discussion

3.1 Force Field Validation

A precise potential that represents atomistic interaction is needed for molecular simulation to be accurate. Before moving on to a more in-depth data analysis from the output of the MD, force field validation is unavoidably important. It should be emphasized that force field validation can occasionally be difficult since some properties are constrained and unable to obtain through an experimental technique. Density is used in this study's comparison because it is possible to access and obtain these data from earlier investigations. The density of the HKUST-1 is displayed in Table 1 where a comparison was conducted between simulations (this work) and experimental data from a prior study. Calculating the percent error revealed a tiny divergence with a 0.00% percent error for density value compared to density obtained by Madden et al. [17]. The force field and simulation settings were briefly confirmed as dependable and trustworthy to generate accurate and fit data because the density value matched the experimental value exactly.

3.2 Solvation-Free Energy and H_2 Adsorption Inside HKUST-1 at Different Temperature

To calculate the excess chemical potential (μ^{ex}) value needed to adsorb hydrogen gas, the solvent-free energy was calculated. The computed value for μ^{ex} at seven different temperatures, ranging from 298 to 363 K, is shown in Table 2. According to Table 2, as temperature values rise, the value of μ^{ex} moves closer to positive values. The gas solubility rate is often determined by the low value of μ^{ex}, with the gases becoming more soluble as μ^{ex} value lowers. At 298 K, the μ^{ex} value is recorded as -0.337 ± 0.001 kcal mol^{-1}, and it rises to -0.292 ± 0.002 kcal mol^{-1} at the

Table 2 Excess chemical potential (μ^{ex}) and Henry's law constant (k_H) inside HKUST-1 at 298–363 K and 1.0 bar

No.	Temperature (K)	μ^{ex} (kcal mol^{-1})	k_H (atm)
1	298	−0.337 ± 0.001	0.986 ± 0.001
2	303	−0.332 ± 0.001	1.020 ± 0.001
3	313	−0.325 ± 0.002	1.085 ± 0.002
4	323	−0.320 ± 0.002	1.146 ± 0.002
5	333	−0.311 ± 0.001	1.217 ± 0.001
6	343	−0.306 ± 0.002	1.280 ± 0.002
7	353	−0.304 ± 0.001	1.338 ± 0.001
8	363	−0.292 ± 0.002	1.416 ± 0.002

final temperature of 363 K. The solubility rate of H_2 in HKUST-1 is reduced even though the μ^{ex} value rises to a positive value. The first foundation for assessing the solubility rate of H_2 in HKUST-1 is the μ^{ex} value. It is evident that the solubility of H_2 decreases as temperature values rise. This pattern matched the findings of a study by Rostami et al. [18], who discovered that H_2 adsorption in MOF-5 dropped rapidly as temperature increased from 296 to 338 K. They claimed that the H_2 absorption is ineffective and falls short of ideal levels at high temperatures. Even Rowsell et al. [19] confirmed that room temperature was unsuitable for the best H_2 absorption. This can be explained by a weakening of the link between hydrogen gas and the adsorbent surface (MOF-5) at high temperatures, which reduces the H_2@HKUST-1 interaction. Goncharov et al. [20] gave the same opinion that the weak physical adsorption of the framework might cause the reduction of hydrogen adsorption at high temperature. Meanwhile, Henry's law constant (k_H) is gradually increased with the decrease of H_2 adsorption. This indicates that the H_2 adsorption calculated in this study is obey the Henry's law where it stated that high k_H value reflect to low gas solubility. The k_H value is calculated to be higher at 363 K with $k_H = 1.416 \pm 0.002$ atm and pose minimum value at 298 K with total $k_H = 0.986 \pm 0.001$ atm.

4 Conclusion

Thermodynamic properties of H_2 adsorption inside HKUST-1 were successfully conducted through solvation-free energy calculation. Based on the first analysis, H_2 molecules showed higher adsorption inside HKUST-1 at low temperatures. At 298 K, the calculated value of μ^{ex} was -0.337 ± 0.001 kcal mol^{-1} and it continued to drop until it was -0.292 ± 0.002 kcal mol^{-1} at 363 K. H_2 adsorption was concluded to decrease with the increasing of the temperature. Henry's law constant showed decreasing in value with increase in H_2 solubility (μ^{ex} value) that actually obeyed Henry's law gas solubility. The general trend stated that higher k_H value reflected to the lower gas solubility as obtained in this study. Overall, the thermodynamic properties were successfully explained and validated since all the analyses showed a good agreement with the previous studies.

Acknowledgements The authors acknowledge the financing support from Yayasan Universiti Teknologi PETRONAS under Graduate Research Assistant Scheme.

References

1. V. Pascanu, G. Gonzalez Miera, A.K. Inge, B. Martin-Matute, Metal-organic frameworks as catalysts for organic synthesis: a critical perspective. J. Am. Chem. Soc. **141**(18), 7223–7234 (2019)
2. H.Y. Li, S.N. Zhao, S.Q. Zang, J. Li, Functional metal-organic frameworks as effective sensors of gases and volatile compounds. Chem. Soc. Rev. **49**(17), 6364–6401 (2020)
3. R. Zhao, Z. Liang, R. Zou, Q. Xu, Metal-organic frameworks for batteries. Joule **2**(11), 2235–2259 (2018)
4. M. Streza, O. Grad, D. Lazar, M. Depriester, S. Longuemart, A.H. Sahraoui, G. Blanita, D. Lupu, Hybrid MOFs-graphene composites: correlation between thermal transport and kinetics of hydrogen adsorption. Int. J. Heat Mass Transfer **143** (2019)
5. B. Han, A. Chakraborty, B.B. Saha, Isosteric heats and entropy of adsorption in Henry's Law region for carbon and MOFs structures for energy conversion applications. Int. J. Heat Mass Transfer **182** (2022)
6. Y. Sun, L. Zheng, Y. Yang, X. Qian, T. Fu, X. Li, Z. Yang, H. Yan, C. Cui, W. Tan, Metal-organic framework nanocarriers for drug delivery in biomedical applications. Nanomicro. Lett. **12**(1), 103 (2020)
7. M.V. Parkes, H. Demir, S.L. Teich-McGoldrick, D.S. Sholl, J.A. Greathouse, M.D. Allendorf, Molecular dynamics simulation of framework flexibility effects on noble gas diffusion in HKUST-1 and ZIF-8. Microporous Mesoporous Mater. **194**, 190–199 (2014)
8. S.S.-Y. Chui, S.M.-F. Lo, J.P.H. Charmant, A.G. Orpen, I.D. Williams, A chemically functionalizable nanoporous material $[Cu_3(TMA)_2(H_2O)_3]_n$. Science **283**(5405), 1148–1150 (1999)
9. N. Greeves, ChemTube3D 2019 responsive version for mobile (Version 3.0) [Computer Software] [Online]. https://www.chemtube3d.com/
10. J.M. MartíNez, L. MartíNez, Packing optimization for automated generation of complex system's initial configurations for molecular dynamics and docking. J. Comput. Chem. **24**(7), 819–825 (2002)
11. M.A.I. Ishak, K. Jumbri, S. Daud, M.B. Abdul Rahman, R. Abdul Wahab, H. Yamagishi, Y. Yamamoto, Molecular simulation on the stability and adsorption properties of choline-based ionic liquids/IRMOF-1 hybrid composite for selective H_2S/CO_2 capture. J. Hazard Mater. **399**, 123008 (2020)
12. G. Garberoglio, OBGMX: a web-based generator of GROMACS topologies for molecular and periodic systems using the universal force field. J. Comput. Chem. **33**(27), 2204–2208 (2012)
13. Y. Chen, Z. Hu, K.M. Gupta, J. Jiang, Ionic liquid/metal–organic framework composite for CO_2 capture: a computational investigation. J. Phys. Chem. C **115**(44), 21736–21742 (2011)
14. U. Essmann, L. Perera, M.L. Berkowitz, T. Darden, H. Lee, L.G. Pedersen, A smooth particle mesh Ewald method. J. Chem. Phys. **103**(19), 8577–8593 (1995)
15. B. Hess, H. Bekker, H.J.C. Berendsen, J.G.E.M. Fraaije, LINCS: a linear constraint solver for molecular simulations. J. Comp. Chem. **18**(12), 1463–1472 (1997)
16. H.J.C. Berendsen, J.P.M. Postma, W.F. van Gunsteren, A. DiNola, J.R. Haak, Molecular dynamics with coupling to an external bath. J. Chem. Phys. **81**(8), 3684–3690 (1984)

17. D.G. Madden, D. O'Nolan, N. Rampal, R. Babu, C. Camur, A.N. Al Shakhs, S.Y. Zhang, G.A. Rance, J. Perez, N.P. Maria Casati, C. Cuadrado-Collados, D. O'Sullivan, N.P. Rice, T. Gennett, P. Parilla, S. Shulda, K.E. Hurst, V. Stavila, M.D. Allendorf, J. Silvestre-Albero, A.C. Forse, N.R. Champness, K.W. Chapman, D. Fairen-Jimenez, Densified HKUST-1 monoliths as a route to high volumetric and gravimetric hydrogen storage capacity. J. Am. Chem. Soc. **144**(30), 13729–13739 (2022)
18. S. Rostami, A. Nakhaei Pour, A. Salimi, A. Abolghasempour, Hydrogen adsorption in metal-organic frameworks (MOFs): effects of adsorbent architecture. Int. J. Hydrogen Energy **43**(14), 7072–7080 (2018)
19. J.L. Rowsell, E.C. Spencer, J. Eckert, J.A. Howard, O.M. Yaghi, Gas adsorption sites in a large-pore metal-organic framework. Science **309**(5739), 1350–1354 (2005)
20. A. Goncharov, A. Guglya, E. Melnikova, On the feasibility of developing hydrogen storages capable of adsorption hydrogen both in its molecular and atomic states. Int. J. Hydrogen Energy **37**, 18061–18073 (2012)

Effect of Temperature on the Hydrogen Adsorption and Transportation Inside MOF-5 Through Molecular Dynamics Simulation

Mohamad Adil Iman Bin Ishak and Khairulazhar Bin Jumbri

Abstract Outstanding and unique properties of MOFs have highlighted their effectiveness in gas storage and transportation especially in H_2 gas adsorption and transportation for green energy conversion. The objective of this study is, to discuss the effect of temperature toward H_2 adsorption inside MOF-5 through molecular dynamics simulation and also, to obtain the atomic properties of the gas adsorption that is not attainable from the experimental work. Solvation-free energy is computed to calculate the adsorption of H_2 inside MOF-5 framework in GROMACS package, followed by Henry's law constant calculation to determine the gas solubility trend based on the gas solubility principle. The outcomes of the analysis will be used as the reference for the future design and modification of MOF-5 for H_2 adsorption and storage.

Keywords Hydrogen adsorption · Metal–organic frameworks · Molecular dynamics simulation

1 Introduction

Hydrogen (H_2) energy is one of the sources that are highly deliberated and commonly spoken to substitute the existing energy sources where it is cheaper, clean, non-toxic, has a high energy capacity and can be produced and reused from various reactions. However, exploration of this energy is very challenging as it requires a deep understanding in terms of hydrogen production, adsorption, storage, transportation, conversion, and reuse. Metal–organic frameworks (MOFs) have emerged as a suitable choice and material for hydrogen adsorption and transportation considering that they have different and special physical characteristics. MOFs are categorized as organic–inorganic hybrid material which is made up of the metal ions that are linked to the organic ligand through coordination bonds to form a uniform crystal-like structure (frameworks). MOFs stand special compared to other adsorbents as they have high

M. A. I. B. Ishak · K. B. Jumbri (✉)
Department of Fundamental and Applied Sciences, Universiti Teknologi PETRONAS, 32610 BandarSeri Iskandar, Perak, Malaysia
e-mail: khairulazhar.jumbri@utp.edu.my

M. B. Othman et al. (eds.), *Proceedings of the 1st International Conference of New Energy*, Springer Proceedings in Energy, https://doi.org/10.1007/978-981-99-0859-2_11

pore volume, high surface area, structural tunability, and high versatility where the selection of metal ion and organic ligand for the bonding is flexible depending on the intended purpose and application. This high nano-porous material are proven to be excellent in various applications such as gas capture [1], adsorption [2], storage [3], transportation [4], drug recovery [5], chemical sensor [6], catalysts [7], battery, and electrolyte [8]. Having a high surface area and large pore volume, IRMOF-1 is chosen as the suitable material for H_2 adsorption and transportation that represents this study. In a similar application, Sillar et al. [9] have conducted a computational study through *ad initio* calculation on the hydrogen adsorption inside MOF-5, where they found that the hydrogen molecules strongly interacted with α-site of the $OZn_4(O_2Ph)_6$ nodes with the interaction energy of $\Delta H_{77} = -7.1$ kJ mol^{-1}. The data also showed that the H_2 interaction was initiated from the local environment around the adsorption site which was ZnO_4. Meanwhile, Wu et al. [10] investigated the effect of temperature on the hydrogen and carbon dioxide adsorption hysteresis in MOF from 77 K until 117 K. For instance, hysteresis is defined as the divergence of the intersection point (contact angle) from its actual (theoretical value) and average value due to physical factor such as microscopic surface defects and roughness [11]. The final data showed that the hysteresis loop in hydrogen adsorption decreased with the increasing of the temperature and completely diminished at 107 and 117 K. From two fitting method applied (Virial and Polynomial), the accurate of hydrogen adsorption were calculated to be at 97.107 and 117 K.

Most of the previous studies were discussed hydrogen adsorption at below room temperature which inadequate justifies the wide range of hydrogen adsorption capacity. Therefore, this study was conducted to determine the effect of temperature (above room temperature) on the hydrogen adsorption inside MOF-5 through Molecular Dynamics (MD) Simulation. H_2-MOF-5 packed molecules were solvated at seven temperatures starting from 303 K until 363 K that give out seven simulated systems in total. The output from simulations was analyzed to calculate the excess chemical potential and Henry's law constant. This information will provide useful guidance toward H_2 adsorption modeling for future improvement.

2 Methodology

The crystal structure of IRMOF-1 was obtained from the online open-access Chem Tube 3D [12]. One subunit, IRMOF-1, is essentially composed of six carboxylate benzene rings acting as the alkyl ligand and four oxide zinc atoms serving as the metal center, as seen in Fig. 1. IRMOF-1 unit cell was built in cubic dimension with a box size of 2.6 × 2.6 × 2.6 nm and placed in the center with ten hydrogen molecules. Both IRMOF-1 and hydrogen molecules were packed together using Packmol [13] installed in the GROMACS package. Hydrogen molecules to be packed were set to ten adopting from the previous study by Ishak et al. [14] and also to ensure appropriate loading since only one subunit IRMOF-1 was used in the simulation.

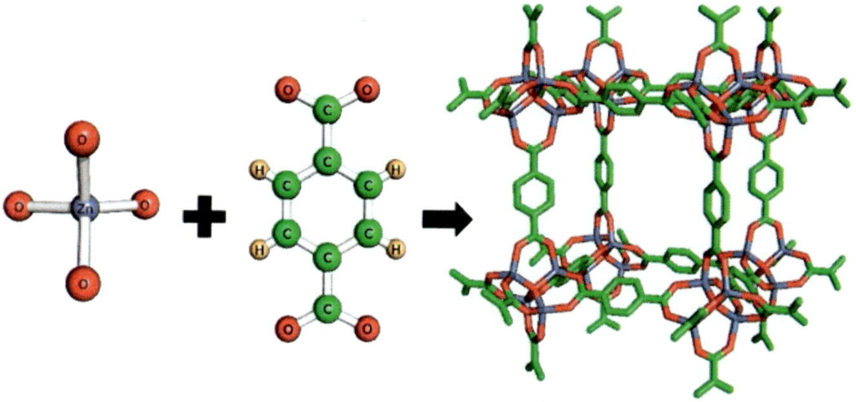

Fig. 1 IRMOF-1 build-up consists of ZnO_4 tetrahedra as metal ions and 1,4-benzenedicarboxylate (BDC) as organic linkers that linked through coordination bonds. Color notation of atom: Zn (gray), Oxygen (red), Carbon (green), Hydrogen (light brown)

All the topologies of hydrogen molecules and IRMOF-1 were generated from the OBGMX [15] topology generator. The generated topology was edited using the force field adopted from Chen et al. [16]. Solvation-free energy was then computed using Bennet Acceptance Ratio (BAR) method to calculate the excess chemical potential. Twenty-one sets of parameters, ranging from 0 to 1, were employed with a total simulation time for each system of five nanoseconds. The non-bonded interactions were calculated up to 1.2 nm, and the long-range electrostatic interactions were handled using Particle-Mesh Ewald (PME) [17] with a grid spacing of 0.12 nm and fourth-order interpolation. The integration step of motion was 2.0 fs. Up to 1.2 nm was covered by the neighbor search, which was updated every five steps. LINCS [18] was used to restrict the bond lengths. The Berendsen thermostat and Berendsen barostat, respectively, were used to managing temperature and pressure [19]. A relaxation time of 2.0 ps and a reference pressure of 1.0 atm was used. The pressure control's isothermal compressibility was set to $4.5 \times 10\text{–}5\ \text{bar}^{-1}$. Two heat baths with temperature coupling constants of 0.1 ps were used to separate the heat. For each steepest descent and conjugate gradient, the systems were started with the energy minimization in 5000 steps. To equilibrate the system at 298 K and 1.0 bar, conical ensemble (NVT) was then applied for a further 100 ps before being followed by the isothermal-isobaric system (NPT) for 5 ns.

Table 1 Density and root-mean squared (RMSD) value of IRMOF-1 from simulated and experimental data at 298 K and 1.0 bar

No	Parameter	Simulated (this work)	Experimental	% Error
1	Density (g/cm^3)	0.61	0.61 [20]	0.00
2	RMSD (Å)	0.20	0.17 [15]	0.15

3 Result and Discussion

3.1 *Force Field Validation*

Accuracy in molecular simulation requires precise potential that depicts atomistic interaction. Force field validation is necessarily important before proceeding to more detailed data analysis from MD output. In this study, density and RMSD (stability) is chosen for the comparison since these data are available to obtain and access from the previous studies. Table 1 shows the density and root mean squared deviation of (RMSD) IRMOF-1 where the comparison was made between simulated (this work) and experimental from the previous study. The percent error was calculated where the comparison showed a small deviation with 0.00% error for density value and 0.15% for RMSD value. Since the density value was the same as the experimental value and the difference in RMSD is lower (less than 1% error), the force field and parameters used in the simulation were briefly validated as reliable and assurance to produce accurate and fit data.

3.2 *Solvation-Free Energy Inside IRMOF-1 at Different Temperatures*

Solvation-free energy was computed to provide excess chemical potential (μ^{ex}) value for H_2 adsorption calculation. Table 2 shows the calculated value for μ^{ex} at seven different temperatures starting from 303 K until 363 K.

Table 2 Excess chemical potential (μ^{ex}) and Henry's law constant (k_H) inside IRMOF-1 at 303–363 K and 1.0 bar

No	Temperature (K)	μ^{ex} (kcal mol^{-1})	k_H (atm)
1	303	−0.191 ± 0.001	1.969 ± 0.001
2	313	−0.189 ± 0.001	2.063 ± 0.001
3	323	−0.186 ± 0.001	2.157 ± 0.001
4	333	−0.184 ± 0.001	2.251 ± 0.001
5	343	−0.182 ± 0.001	2.346 ± 0.001
6	353	−0.179 ± 0.001	2.441 ± 0.001
7	363	−0.174 ± 0.001	2.545 ± 0.001

Based on Table 2, the value of μ^{ex} is increasingly approaching positive values with the increasing temperature values. Typically, the low value of μ^{ex} determines the gas solubility rate, where the gases become more soluble as the μ^{ex} value decreases further. The μ^{ex} value recorded at a temperature of 303 K is −0.191 ± 0.001 kcal mol^{-1} and continues to increase to −0.174 ± 0.001 kcal mol^{-1} at the final temperature of 363 K. Although the μ^{ex} value increases to a positive value, it shows a reduction in the solubility rate of H_2 in IRMOF-1. It can be concluded that H_2 solubility is decreasing with increasing temperature values. This trend was similar to a study done by Rostami et al. [21] where they found that H_2 adsorption in MOF-5 decreased rapidly with the increasing of temperature starting from 296 to 338 K. They stated that the H_2 absorption is inappropriate at high temperatures and does not reach the optimum values. Even the study conducted by Rowsell et al. [22] also shows that room temperature is not suitable for optimal H_2 absorption. This can be explained through a bond strength between hydrogen gas and adsorbent surface (MOF-5) [23, 24] where at high temperatures the bond between them will be weakened, reducing the H_2@IRMOF-1 interaction. Henry's law (k_H) calculation shows a good correlation with the H_2 solubility where the calculated data in this study obeys Henry's law of gas solubility. It satisfies the relationship where the k_H values increased with the decreasing of H_2 solubility. High k_H value is observed at 363 K with calculated $k_H = 2.545 \pm 0.001$ atm but eventually reduced at low temperature with minimum k_H value of 1.969 ± 0.001 atm at 303 K.

4 Conclusion

The effect of the temperature on the hydrogen adsorption inside IRMOF-1 was successfully conducted. At the end of this study, the H_2 adsorption is greatly influenced by the temperature exerted to the system. The H_2 adsorption was quantitatively measured where at 303 K, the μ^{ex} recorded was −0.191 ± 0.001 kcal mol^{-1} and continued to decrease to −0.174 ± 0.001 001 kcal mol^{-1} at the temperature of 363 K. It is proven that all the data calculated in this study is reliable since the H_2 solubility calculated obeys Henry's law of gas solubility. The final trend follows the principle where the H_2 solubility decreases with the increase of the k_H values. This available observation is actually useful for the future development and modification of the H_2 adsorption specifically toward MOFs creation.

Acknowledgements Authors and members would like to thank Yayasan Universiti Teknologi PETRONAS for research grant funding (015LC0-211).

References

1. X. Zhang, Q.-R. Zheng, H.-Z. He, Machine-learning-based prediction of hydrogen adsorption capacity at varied temperatures and pressures for MOFs adsorbents. J. Taiwan Inst. Chem. Eng. **138** (2022)
2. B. Han, A. Chakraborty, B.B. Saha, Isosteric heats and entropy of adsorption in Henry's law region for carbon and MOFs structures for energy conversion applications. Int. J. Heat Mass Transf. **182** (2022)
3. K. Koizumi, K. Nobusada, M. Boero, Hydrogen storage mechanism and diffusion in metal-organic frameworks. Phys. Chem. Chem. Phys. **21**(15), 7756–7764 (2019)
4. E. Aliyev, J. Warfsmann, B. Tokay, S. Shishatskiy, Y.-J. Lee, J. Lillepaerg, N.R. Champness, V. Filiz, Gas transport properties of the metal-organic framework (MOF)-assisted polymer of intrinsic microporosity (PIM-1) thin-film composite membranes. ACS Sustain. Chem. Eng. **9**(2), 684–694 (2020)
5. Y. Sun, L. Zheng, Y. Yang, X. Qian, T. Fu, X. Li, Z. Yang, H. Yan, C. Cui, W. Tan, Metal-organic framework nanocarriers for drug delivery in biomedical applications. Nanomicro Lett. **12**(1), 103 (2020)
6. H.Y. Li, S.N. Zhao, S.Q. Zang, J. Li, Functional metal-organic frameworks as effective sensors of gases and volatile compounds. Chem. Soc. Rev. **49**(17), 6364–6401 (2020)
7. I. Abanades Lazaro, R.S. Forgan, F.G. Cirujano, MOF nanoparticles as heterogeneous catalysts for direct amide bond formations. Dalton Trans. **51**(21), 8368–8376 (2022)
8. R. Zhao, Z. Liang, R. Zou, Q. Xu, Metal-organic frameworks for batteries. Joule **2**(11), 2235–2259 (2018)
9. K. Sillar, A. Hofmann, J. Sauer, Ab Initio study of hydrogen adsorption in MOF-5. J. Am. Chem. Soc. **131**, 4143–4150 (2009)
10. H. Wu, C.G. Thibault, H. Wang, K.A. Cychosz, M. Thommes, J. Li, Effect of temperature on hydrogen and carbon dioxide adsorption hysteresis in an ultramicroporous MOF. Micropor. Mesopor. Mater. **219**, 186–189 (2016)
11. J. Berthier, Electrowetting theory, in *Micro-Drops and Digital Microfluidics*, 2nd edn. (William Andrew Publishing, 2013), pp. 161–224
12. N. Greeves, ChemTube3D 2019 responsive version for mobile (Version 3.0) [Computer Software] [Online]. https://www.chemtube3d.com/
13. J. M. Martínez, L. Martínez, Packing optimization for automated generation of complex system's initial configurations for molecular dynamics and docking. J. Comput. Chem. **24**(7), 819–825 (2002)
14. M.A.I. Ishak, K. Jumbri, S. Daud, M.B. Abdul Rahman, R. Abdul Wahab, H. Yamagishi, Y. Yamamoto, Molecular simulation on the stability and adsorption properties of choline-based ionic liquids/IRMOF-1 hybrid composite for selective H2S/CO2 capture. J. Hazard. Mater. **399**, 123008 (2020)
15. G. Garberoglio, OBGMX: a web-based generator of GROMACS topologies for molecular and periodic systems using the universal force field. J. Comput. Chem. **33**(27), 2204–2208 (2012)
16. Y. Chen, Z. Hu, K.M. Gupta, J. Jiang, Ionic liquid/metal–organic framework composite for CO2 capture: a computational investigation. J. Phys. Chem. C **115**(44), 21736–21742 (2011)
17. T. Darden, D. York, L. Pedersen, Particle mesh Ewald: an N·log (N) method for Ewald sums in large systems. J. Chem. Phys. **98**(12), 10089–10092 (1993)
18. B. Hess, H. Bekker, H.J.C. Berendsen, J.G.E.M. Fraaije, LINCS: a linear constraint solver for molecular simulations. J. Comp. Chem. **18**(12), 1463–1472 (1997)
19. H.J.C. Berendsen, J.P.M. Postma, W.F. van Gunsteren, A. DiNola, J.R. Haak, Molecular dynamics with coupling to an external bath. J. Chem. Phys. **81**(8), 3684–3690 (1984)
20. M. Eddaoudi, J. Kim, N. Rosi, D. Vodak, J. Wachter, M. O'Keeffe, O.M. Yaghi, Systematic design of pore size and functionality in isoreticular mofs and their application in methane storage. Science **295**(5554), 469–472 (2002)

21. S. Rostami, A. Nakhaei Pour, A. Salimi, A. Abolghasempour, Hydrogen adsorption in metal-organic frameworks (MOFs): effects of adsorbent architecture. Int. J. Hydrog. Energy **43**(14), 7072–7080 (2018)
22. J.L. Rowsell, E.C. Spencer, J. Eckert, J.A. Howard, O.M. Yaghi, Gas adsorption sites in a large-pore metal-organic framework. Science **309**(5739), 1350–1354 (2005)
23. W. Zhao, V. Fierro, C. Zlotea, M.T. Izquierdo, C. Chevalier-César, M. Latroche, A. Celzard, Activated carbons doped with Pd nanoparticles for hydrogen storage. Int. J. Hydrog. Energy **37**(6), 5072–5080 (2012)
24. M.C. Tellez-Juárez, V. Fierro, W. Zhao, N. Fernández-Huerta, M.T. Izquierdo, E. Reguera, A. Celzard, Hydrogen storage in activated carbons produced from coals of different ranks: effect of oxygen content. Int. J. Hydrog. Energy **39**(10), 4996–5002 (2014)

A Review on Bio-hydrogen Production from Food Waste: Potential and Challenges

Md. Sohrab Hossain, Md. Mokarram Badsha, Venugopal Balakrishnan, and Maizatul Shima Shaharun

Abstract The global energy demand is increasing immensely with rapid population growth, urbanization, and industrialization. At present, fossil fuel is the major source of world energy. However, the rapid depletion of fossil fuel energy, environmental pollution concerns, and increasing demand due to rapid population growth, urbanization, and industrialization have led to exploring alternative energy sources to conventional fossil fuel. Bio-hydrogen is viewed as a promising renewable biofuel and it has great potential to mitigate the rising energy demand. Due to the great energy potential of hydrogen, a number of studies have been conducted on hydrogen production from various renewable resources. Challenges that are being faced in large-scale hydrogen production are low hydrogen yield, selection of suitable technology, and availability of feedstock for hydrogen production. A large amount of food waste is generated every day. The most common disposal method of food waste is open dumping in a landfill. This inappropriate disposal of food waste poses severe environmental pollution. However, food waste could be utilized as a potential feedstock for bio-hydrogen production because of its significant amount of carbohydrate, fate, and protein compositions. Therefore, the present study was conducted to determine the potential and limitations of utilizing food waste as a source for bio-hydrogen production by reviewing the compositions of food waste, and factors

Md. Sohrab Hossain (✉) · M. S. Shaharun
HICoE-Centre for Biofuel and Biochemical Research, Institute of Self-Sustainable Building, Department of Fundamental and Applied Sciences, Universiti Teknologi PETRONAS, 32610 Seri Iskandar, Perak Darul Ridzuan, Malaysia
e-mail: sohrab.hossain@utp.edu.my

Md. Sohrab Hossain · Md. Mokarram Badsha
University of Kuala Lumpur-Malaysian Institute Chemical & Bioengineering Technology (UniKL-MICET), 78000 Alor Gajah, Melaka, Malaysia

V. Balakrishnan
Institute for Research in Molecular Medicine (INFORMM), Universiti Sains Malaysia, 11800 USM, George Town, Penang, Malaysia

M. S. Shaharun
Institute of Contaminant Management, Centre for Contaminant Control & Utilization (CenCoU), Department of Fundamental and Applied Sciences, Universiti Teknologi PETRONAS, 32610, Seri Iskandar, Malaysia

M. B. Othman et al. (eds.), *Proceedings of the 1st International Conference of New Energy*, Springer Proceedings in Energy, https://doi.org/10.1007/978-981-99-0859-2_12

influencing hydrogen production from food waste. The findings of the present study will be a harmony of sustainable utilization of food waste as a potential source for bio-hydrogen production.

Keywords Hydrogen · Food waste · Renewable energy · Sustainability

1 Introduction

Food waste is discarded and destroyed food products from restaurants, household kitchens, cafeterias, or food processing plants. Approximately, one-third of the produced foods worldwide discards as food waste. It is estimated that about 1.3 billion tons of food waste are generated annually. However, it is expected that food waste generation will increase massively with the population and economic growth. The safe disposal of food waste poses a significant challenge worldwide due to the enormous volume of generation, suitable disposal technology, human health, and environmental pollution concerns [1, 2]. Although food waste is biodegradable and not considered hazardous waste, the inappropriate treatment and disposal of food waste may cause severe environmental pollution, foul odor, and greenhouse gas emissions [2, 3]. The most common technology for the disposal of food waste is dumping the waste in a landfill. Generally, food waste is disposed of in landfill along with other municipal solid waste. Although landfilling of food waste is the most convenient disposal method because of its inherent advantages, including large disposal capacity, ease of operation and management, low capital investment, and low maintenance cost [3]. However, the landfilling of food waste has some inevitable limitations such as the generation of large volumes of landfill leachate, a large carbon footprint, poor sanitation, and the generation of greenhouse gasses [2, 4].

Food waste mainly consists of carbohydrates, proteins, lipids, and trace amounts of inorganic compounds. However, the composition of the food waste may vary with its constituents. For example, food waste comprising rice, wheat, and vegetables will be rich in carbohydrates. Conversely, the food waste containing eggs and meat will be rich in lipids and protein [1, 5]. Studies reported that food waste contains about 40 wt% of carbohydrates, 40 wt% of lipids, and 20 wt% of protein [6]. Thus, food waste can be utilized as an ideal substrate for bioenergy, bio-fertilizer, and other value-added product production. Over the years, various biological methods such as anaerobic digestion, fermentation, and composting have been implemented to produce bio-methane [7], bio-ethanol [8], and bio-fertilizer [9] production from food waste. However, scientists and environmentalists are facing numerous obstacles in the production of bio-methane, bio-ethanol, and bio-fertilizer with the imposed biological methods, including high operating cost, high process complexity, large footprint, and post-handling of residue [6, 7, 9]. Therefore, it is urged for an efficient technology to conduct effective food waste management with low operating cost, marginal solid residue generation, resource recovery, and a small carbon footprint.

There is an increasing interest in determining alternative energy sources due to the rapid depletion of fossil fuel-based energy. The energy depletion concern is alarming with the rapid population growth, industrialization, and urbanization [1]. Hydrogen is an efficient and clean fuel, since it just produces water as a by-product during production [7, 10]. Hydrogen is viewed as a promising alternative energy source of conventional fossil fuel energy because of its 2.75 times higher energy content than fossil fuel energy. Thus, there is ample opportunity to compete with rising energy demand and depletion of fossil fuel energy with the production of hydrogen. Besides, the production and utilization of hydrogen fuel will minimize air pollution and combat global climate change. The major challenges of producing hydrogen are the feed-stock price and suitable conversion technology. At present, over 96% of hydrogen is produced from fossil fuel sources [7, 11]. However, hydrogen can be produced from carbohydrate, protein, and lipids-rich substrates using a thermochemical or fermentative conversion process. Since food waste is enriched with carbohydrates, protein, and lipids, therefore food waste could be utilized as a potential and cheap substrate for hydrogen production. It is, therefore, the present study was aim to review the feasibility of utilizing food waste as feedstock for hydrogen production. Besides, the implementation of suitable technology for the production of hydrogen from food waste, its advantages, and limitations are also discussed in the present study.

2 Compositions of the Food Waste Toward Hydrogen Production

Food waste is a type of organic waste generated during food processing, discarded and unwanted fruits or food products in food processing industries, restaurants, cafeterias, and household kitchens. The composition of food waste may vary with various factors, including source, food supply chain, food habits, and economic levels [2, 6]. However, the food waste generated, particularly, in Asian countries, are rich in carbohydrate, protein, and fats [4]. Table 1 presents the percentage of carbohydrate, fats, and protein content in various types of food waste.

2.1 Carbohydrate

Studies reported that food waste enriched in carbohydrate content could be utilized as an ideal substrate for hydrogen production [12]. This is because carbohydrate is easily biodegradable and therefore carbohydrate turn to hydrogen gas by the digestion of fermentative microorganisms [11]. The mechanisms of hydrogen production from carbohydrate-rich food waste can be divided into acetate and butyrate pathways, as shown below.

Table 1 Compositions of the various types of food waste toward hydrogen production

Food waste	[a]Compositions (wt%)				References
	Carbohydrate	Protein	Lipids	Others	
Kitchen waste	28.4	21.6	19.4	21.6	[10]
Kitchen waste	34.7	29.3	22.3	20.8	[11]
Kitchen waste	41.32	16.71	12.66		[12]
Waste Egg		66.6	33.5		[10]
Waste cooking oil	5.3	0.2	94.5		[10]
Food waste	39.1	13.7	12.3	5.2	[13]
Watermelon	80.4	7.4	4	8.2	[14]
Melon	75.7	10.2	4.8	9.3	[14]
Kitchen waste	40.2	16.3	–	5.7	[15]
Food waste	33.3	10.4	15.3	NA	[16]
Bakery waste	62.1	4.2	19.0	NA	[16]
Food waste	39.5	11			[17]
Organic waste	57.7	23.9	12.8	5.6	[18]
Food waste	36.0	41.5	18.5	NA	[19]

[a] dry basis

$$C_6H_{12}O_6 + 2H_2O \rightarrow 2CH_3COOH + 2CO_2 + 4H_2 \quad (1)$$

$$C_6H_{12}O_6 + 2H_2O \rightarrow CH_3CH_2CH_2COOH + 2CO_2 + 2H_2 \quad (2)$$

$$C_6H_{12}O_6 + 2H_2O \rightarrow CH_3COOH + CH_3CH_2OH + 2CO_2 + 2H_2 \quad (3)$$

$$C_6H_{12}O_6 + 2H_2O \rightarrow C_3H_6O + 3CO_2 + 4H_2 \quad (4)$$

$$4C_6H_{12}O_6 + 2H_2O \rightarrow 3CH_3CH_2CH_2COOH + 2CH_3COOH + 8CO_2 + 10H_2 \quad (5)$$

Jarunglumlert et al. [20] reported that food waste with high amounts of carbohydrates, such as starch and simple sugar has the potential to be utilized as a promising substrate for hydrogen production. Similarly, Guo et al. [21] determined the hydrogen production potential of 26 different types of organic solid waste substrates. They found that the solid organic waste containing the highest amount of carbohydrates produced the maximum amount of hydrogen, and protein-rich substrate produced the lowest amount of hydrogen gas. Generally, carbohydrate-rich substrate produces a higher amount of hydrogen than protein and fate-rich substrates [16, 22]. For instance, Lay et al. [23] conveyed that carbohydrate-rich substrates produce 20 times higher hydrogen gas than protein-rich substrates, and 16 times higher than fat-rich substrates. Alibardi et al. [24] determine the potential of hydrogen production of 4 different types of food waste, including (i) carbohydrate-rich food waste—a mixture of pasta and bread, (ii) protein and the fat-rich substrates—a mixture of cheese, fish, and meat, (iii) vegetable and (iv) fruits. It was observed that the carbohydrate-rich substrate produced the highest amount of hydrogen gas (152–167 H_2/gVS). Wherein, the vegetable and fruits yielded 149–188 H_2/gVS, and protein-rich substrate produced 0.8–5 H_2/gVS. The statistical study of this revealed that hydrogen production significantly correlated with the carbohydrate content of the substrate, wherein protein, lipids, and C/N ratio of the substrate showed an insignificant effect on hydrogen production. Although, carbohydrate is found the ideal substrate for bio-hydrogen production, cellulose base carbohydrate or complex carbohydrate produces a minimal amount of hydrogen [22, 25]. The complex cross-link between the polysaccharides of cellulosic materials, such as lignin, cellulose, and hemicellulose inhibits hydrogen production, because the applied hydrogen production technology is unable to break down the large molecules. Therefore, the cellulosic materials require pre-treatment such as hydrolysis to break down the large polysaccharide molecules into smaller molecules of mono or polysaccharides [20].

2.2 *Fats and Lipids*

Food waste contains about 13–16% of lipids or fats. The source of the lipids in food waste is dairy products, vegetable oil, and animal and fish fats [20]. Studies reported that lipids or fats are not the sole substrates for hydrogen production [20, 26]. The presence of fats and lipids in food waste causes mass transfer problems because of floating or clogging the fermenter during hydrogen production using the microbial fermentation process [26]. Therefore, the food waste must be pre-treated with a hydrothermal pre-treatment process to degrade the fats to avoid clogging the fermenter during hydrogen production. Numerous studies reported that the fats and lipids composition in food waste offers minimal hydrogen production. Fats or lipids-rich food waste is not an ideal substrate for hydrogen production when biological fermentation process is used [22]. However, the presence of fats in the food waste enhances the carbohydrate fermentation rate and therefore increases hydrogen production [20]. The mechanisms of hydrogen production from fats and lipids-rich food waste using the fermentation process can be described using Eq. (6).

$$n\mathrm{LCFA} \rightarrow (n-2)\mathrm{LCFA} + 2\mathrm{CH_3COO^-} + 2\mathrm{H_2} \tag{6}$$

where lipids and fats present in the food waste hydrolyze with microorganisms enzyme and produce long-chain fatty acids (LCFA). The process protects the nutrient adsorption of hydrogen-consuming microbes during microbial fermentation and therefore it reduces the number of hydrogen-consuming microbes and enhances the carbohydrate degradation rate to produce hydrogen [21, 26].

2.3 Protein

Proteins are macromolecules and biomolecules that comprise one or more long-chain peptides or amino acids. The source of protein in food waste is from discarded protein-rich foods, including fish, eggs, meat, cheese, casein, and whey [26]. However, the presence of protein in the food waste enhances the nutrient source for the hydrogen-producing bacteria [14, 27]. Therefore, the presence of protein in food waste enhances hydrogen production from food waste [27]. During the anaerobic digestion of food waste, the protease enzymes are produced by hydrogen-producing bacteria. Subsequently, the produced enzyme hydrolyze protein to produce amino acid. The amino acids further breakdown into fatty acids, ammonia, hydrogen, and carbon dioxide gas [26, 27]. For example, Cheng et al. [28] reported that hydrogen was produced from serine and alanine by digesting with *Clostridium* bacteria in a dark fermentation process by the following pathways:

$$C_3H_{17}NO_3 + H_2O \xrightarrow{\textit{Clostridium}\ \text{sp.}} CH_3COOH + CO_2 + NH_3 + H_2 \quad (7)$$

$$3C_3H_7NO_2 + 3H_2O \xrightarrow{\textit{Clostridium}\ \text{sp.}} CH_3COOH + 2CH_3CH_2COOH + CO_2 + 3NH_3 + H_2 \quad (8)$$

$$2C_3H_7NO_2 + 2H_2O \xrightarrow{\textit{Clostridium}\ \text{sp.}} CH_3CH_2CH_2COOH + 2CO_2 + 2NH_3 + 2H_2 \quad (9)$$

Studies reported that protein-rich food waste is not a good substrate for hydrogen production due to its slow anaerobic degradation [26]. The complex and long-chain structure of protein caused difficulties in breakdown by microorganisms and therefore it slows hydrogen production. However, the production of hydrogen can be improved from protein-rich food waste by adding a hydrothermal pre-treatment process of food waste to break down protein into simple amino acids [20, 24].

3 Advantages and Limitations of H_2 Production from Food Waste

There is increasing interest in the sustainable utilization of food waste as a renewable feedstock for bio-energy production due to its enormous volume of generation, suitable compositions for bioenergy, and minimizing environmental pollution and greenhouse gas emission [4]. The utilization of food waste as the feedstock of renewable energy, not only reduces the volume of food waste disposed of in a landfill, it also generates capital from this unwanted food waste. Figure 1 shows the existing hydrogen production from various renewable and renewable resources like

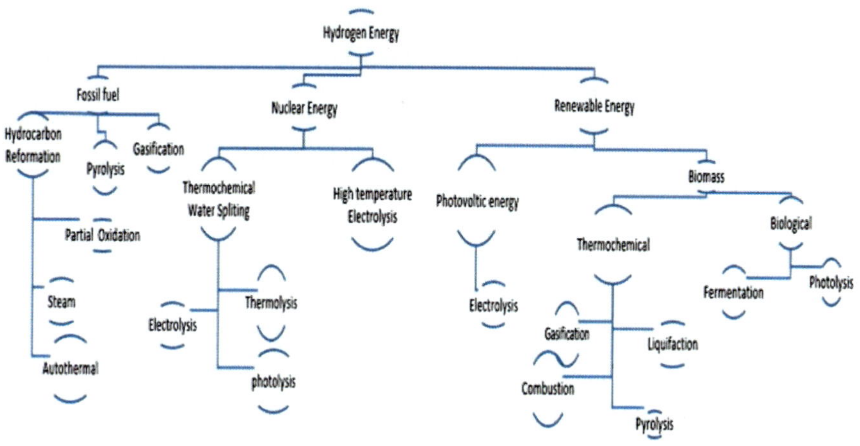

Fig. 1 Hydrogen production technologies from renewable and non-renewable feedstock

biomass [20, 22, 26]. The hydrogen can be produced from biomass using either thermochemical or biological processes [22]. However, food waste contains high levels of moisture, and therefore the thermochemical process is not suitable for hydrogen production from the food waste. In recent years, fermentation technology has been viewed as the most preferred technology for hydrogen production from biomass due its simplicity and easiest biological method of hydrogen production [27]. The fermentation process could be utilized to produce hydrogen from food waste. The implementation of the fermentation technology for hydrogen production from food waste would enhance the sustainable utilization of the waste materials. The major limitation of the existing hydrogen production technologies from various biomass including food processing waste is low hydrogen yield, and low substrate conversion [26]. However, hydrogen production efficiency can be improved with the application of integrated hybrid fermentation technology. Several studies in literature have implemented hybrid fermentation technology along with bio-electrochemical technology, photo-fermentation, and anaerobic digestion method [22, 29].

The major limitations of large-scale bio-hydrogen production include the limitation of the reliable substrate, selection of suitable technology, and the cost of hydrogen production [20, 29]. At present, fossil fuel is the major source of hydrogen energy production. However, the utilization of renewable resources, such as solar energy, wind, and biomass in recent years gradually minimizes the dependency on fossil fuels [16]. Food waste because of its significant carbohydrate content could be utilized as a potential substrate for hydrogen production. Numerous studies have been conducted on the production of hydrogen from food processing waste and food waste using dark fermentation technology [12, 18, 26, 28]. The studies reported that the utilization of food waste as a substrate would minimize hydrogen energy production cost, enhance sustainable utilization of solid waste, and minimize greenhouse gas emissions. The major challenge of utilizing food waste as a substrate for hydrogen production is the high moisture content, presence of cellulosic materials, and protein and fat compositions. Although, dark fermentation is found to be a suitable technology for hydrogen production from food waste, the presence of cellulosic materials, fats, and protein compositions may require a pre-treatment process of food waste to enhance hydrogen yield. Therefore, additional research is required to determine a suitable pre-treatment process for food waste and to assess the techno-economic viability of hydrogen production from food waste.

4 Conclusions

In the present study, the potential of utilizing food waste as a feedstock for hydrogen production was reviewed. It was found that food waste contains significant amounts of carbohydrates, fats, and proteins. However, the carbohydrate content in food waste is higher than other components, thus the food waste could be utilized as a potential substrate for hydrogen production. However, fermentation, in particular, dark fermentation is a suitable technology for the production of hydrogen from food waste. Since food waste contains fats, proteins, and cellulosic materials, therefore it urges for a pre-treatment process to enhance the hydrogen production from food waste. The finding of the present study reveals that food waste could be utilized as a potential feedstock for hydrogen production. However, it urges additional studies to determine a suitable pre-treatment process and assess the techno-economic viability of hydrogen production from food waste.

Acknowledgements The authors acknowledge the financial support by the HICoE-Centre for Biofuel and Biochemical Research, Institute of Self-Sustainable Building, Universiti Teknologi PETRONAS.

References

1. T.P.T. Pham, R. Kaushik, G.K. Parshetti, R. Mahmood, R. Balasubramanian, Food waste-to-energy conversion technologies: current status and future directions. Waste Manag. **38**, 399–408 (2015)
2. H.B. Sharma, K.R. Vanapalli, V.R.S. Cheela, V.P. Ranjan, A.K. Jaglan, B. Dubey, S. Goel, J. Bhattacharya, Challenges, opportunities, and innovations for effective solid waste management during and post COVID-19 pandemic. Resour. Conserv. Recycl. **162**, 105052 (2020)
3. S. Sinha, P. Tripathi, Trends and challenges in valorisation of food waste in developing economies: a case study of India. Case Stud. Chem. Environ. Eng. **4**, 100162 (2021)
4. S.L. Nordahl, J.P. Devkota, J. Amirebrahimi, S.J. Smith, H.M. Breunig, C.V. Preble, A.J. Satchwell, L. Jin, N.J. Brown, T.W. Kirchstetter, C.D. Scown, Life-cycle greenhouse gas emissions and human health trade-offs of organic waste management strategies. Environ. Sci. Technol. **54**(15), 9200–9209 (2020)

5. H. Su, M. Yan, S. Wang, Recent advances in supercritical water gasification of biowaste catalyzed by transition metal-based catalysts for hydrogen production. Renew. Sustain. Energy Rev. **154**, 111831 (2022)
6. L. Yue, J. Cheng, J. Hua, H. Dong, J. Zhou, Y.-Y. Li, Improving fermentative methane production of glycerol trioleate and food waste pretreated with ozone through two-stage dark hydrogen fermentation and anaerobic digestion. Energy Convers. Manag. **203**, 112225 (2020)
7. D.E. Algapani, W. Qiao, M. Ricci, D. Bianchi, S.M. Wandera, F. Adani, R. Dong, Bio-hydrogen and bio-methane production from food waste in a two-stage anaerobic digestion process with digestate recirculation. Renew. Energy **130**, 1108–1115 (2019)
8. D. Mathioudakis, P. Karageorgis, K. Papadopoulou, T.F. Astrup, G. Lyberatos, Environmental and economic assessment of alternative food waste management scenarios. Sustainability **14**(15), 9634 (2022)
9. L. Wang, Y. Chi, K. Du, Z. Zhou, F. Wang, Q. Huang, Hydrothermal treatment of food waste for bio-fertilizer production: formation and regulation of humus substances in hydrochar. Sci. Total Environ. **838**, 155900 (2022)
10. T. Kobayashi, K.-Q. Xu, Y.-Y. Li, Y. Inamori, Evaluation of hydrogen and methane production from municipal solid wastes with different compositions of fat, protein, cellulosic materials and the other carbohydrates. Int. J. Hydrogen Energy **37**(20), 15711–15718 (2012)
11. F. Wang, C. Zhou, W. He, H. Zhu, J. Huang, G. Li, The content variation of fat, protein and starch in kitchen waste under microwave radiation. Procedia Environ. Sci. **31**, 530–534 (2016)
12. L. Xin, X. Yan, X. Xu, Y. Qin, Q. Nan, H. Wang, W. Wu, Carbohydrate degradation contributes to the main bioheat generation during kitchen waste biodrying process: a pilot study. Waste Manag. **137**, 20–30 (2022)
13. G. Prasoulas, A. Gentikis, A. Konti, S. Kalantzi, D. Kekos, D. Mamma, Bioethanol production from food waste applying the multienzyme system produced on-site by fusarium oxysporum F3 and mixed microbial cultures. Fermentation **6**(2), 39 (2020)
14. S. Greses, E. Tomás-Pejó, C. González-Fernández, Short-chain fatty acids and hydrogen production in one single anaerobic fermentation stage using carbohydrate-rich food waste. J. Clean. Prod. **284**, 124727 (2021)
15. H. Nishimura, L. Tan, N. Kira, S. Tomiyama, K. Yamada, Z.-Y. Sun, Y.-Q. Tang, S. Morimura, K. Kida, Production of ethanol from a mixture of waste paper and kitchen waste via a process of successive liquefaction, presaccharification, and simultaneous saccharification and fermentation. Waste Manag. **67**, 86–94 (2017)
16. D. Pleissner, W.C. Lam, Z. Sun, C.S.K. Lin, Food waste as nutrient source in heterotrophic microalgae cultivation. Biores. Technol. **137**, 139–146 (2013)
17. K. Wang, J. Yin, D. Shen, N. Li, Anaerobic digestion of food waste for volatile fatty acids (VFAs) production with different types of inoculum: effect of pH. Biores. Technol. **161**, 395–401 (2014)
18. X. Lu, W. Jin, S. Xue, X. Wang, Effects of waste sources on performance of anaerobic co-digestion of complex organic wastes: taking food waste as an example. Sci. Rep. **7**(1), 15702 (2017)
19. Y. Li, Y. Jin, H. Li, A. Borrion, Z. Yu, J. Li, Kinetic studies on organic degradation and its impacts on improving methane production during anaerobic digestion of food waste. Appl. Energy **213**, 136–147 (2018)
20. T. Jarunglumlert, C. Prommuak, N. Putmai, P. Pavasant, Scaling-up bio-hydrogen production from food waste: feasibilities and challenges. Int. J. Hydrogen Energy **43**(2), 634–648 (2018)
21. X.M. Guo, E. Trably, E. Latrille, H. Carrere, J.-P. Steyer, Predictive and explicative models of fermentative hydrogen production from solid organic waste: role of butyrate and lactate pathways. Int. J. Hydrogen Energy **39**(14), 7476–7485 (2014)
22. M. Kaya, Evaluating organic waste sources (spent coffee ground) as metal-free catalyst for hydrogen generation by the methanolysis of sodium borohydride. Int. J. Hydrogen Energy **45**(23), 12743–12754 (2020)
23. J.-J. Lay, K.-S. Fan, J. Chang, C.-H. Ku, Influence of chemical nature of organic wastes on their conversion to hydrogen by heat-shock digested sludge. Int. J. Hydrogen Energy **28**(12), 1361–1367 (2003)

24. L. Alibardi, R. Cossu, Effects of carbohydrate, protein and lipid content of organic waste on hydrogen production and fermentation products. Waste Manag. **47**, 69–77 (2016)
25. S. Chozhavendhan, M. Rajamehala, G. Karthigadevi, R. Praveenkumar, B. Bharathiraja, A review on feedstock, pretreatment methods, influencing factors, production and purification processes of bio-hydrogen production. Case Stud. Chem. Environ. Eng. **2**, 100038 (2020)
26. N.H.M. Yasin, T. Mumtaz, M.A. Hassan, N.A. Abd Rahman, Food waste and food processing waste for biohydrogen production: a review. J. Environ. Manag. **130**, 375–385 (2013)
27. G.K. Dinesh, R. Chauhan, S. Chakma, Influence and strategies for enhanced biohydrogen production from food waste. Renew. Sustain. Energy Rev. **92**, 807–822 (2018)
28. J. Cheng, L. Ding, A. Xia, R. Lin, Y. Li, J. Zhou, K. Cen, Hydrogen production using amino acids obtained by protein degradation in waste biomass by combined dark- and photo-fermentation. Biores. Technol. **179**, 13–19 (2015)
29. J. Baeyens, H. Zhang, J. Nie, L. Appels, R. Dewil, R. Ansart, Y. Deng, Reviewing the potential of bio-hydrogen production by fermentation. Renew. Sustain. Energy Rev. **131**, 110023 (2020)

Facile Fabrication of PTA@MOF-808H Nanocomposites in Acidic Media Employing Hydrogen Peroxide for Catalytic Oxidative Desulfurization of Fuel Oil

Abdurrashid Haruna, Zulkifli Merican Aljunid Merican, and Suleiman Gani Musa

Abstract This research paper describes the fabrication of a series of PTA@MOF-808(Zr)H composite synthesized by the impregnation method. The structural and microstructural properties of the nanocomposites were investigated by using Fourier transform infrared, Thermogravimetric analysis, Powdered x-ray diffraction, and Field emission scanning electron microscopy. The structural characterization indicated that PTA was successfully incorporated into the cavity of MOF-808(Zr) while the structure could remain stable at 350 °C. For this purpose, the produced catalysts were used for the ODS reaction of model fuel injected with BT, DBT, and 4,6-DMBT. The results showed that the highest conversion of DBT is obtained using PTA@MOF-808(Zr)-0.5H as an efficient catalyst in acidic media employing hydrogen peroxide. An improvement in the conversion rate of DBT was achieved at nearly 100% efficiency within a short time. Besides, the catalyst demonstrated high conversion efficiency and therefore may be potentially used for hydrogen generation and storage.

Keywords Metal–organic frameworks · MOF-808 · POM@MOF · Porous materials · Oxidative desulfurization · Fuel oil

A. Haruna · Z. M. A. Merican (✉) · S. G. Musa
Department of Fundamental and Applied Sciences, Universiti Teknologi PETRONAS, 32610 Seri Iskandar, Perak Darul Ridzuan, Malaysia
e-mail: zulkifli.aljunid@utp.edu.my

A. Haruna
Department of Chemistry, Ahmadu Bello University, Zaria, Nigeria

A. Haruna · Z. M. A. Merican
Institute of Contaminant Management, Universiti Teknologi PETRONAS, 32610 Seri Iskandar, Perak Darul Ridzuan, Malaysia

S. G. Musa
Department of Chemistry, Al-Qalam University Katsina, Tafawa Balewa Way, Katsina, Nigeria

M. B. Othman et al. (eds.), *Proceedings of the 1st International Conference of New Energy*, Springer Proceedings in Energy, https://doi.org/10.1007/978-981-99-0859-2_13

1 Introduction

More than 80% of energy comes from petroleum, coal, and natural gas and about 98% of environmental pollution comes from the combustion of fossil fuels. Reducing the use of fuel containing sulfur compounds would considerably reduce the impacts of environmental pollutants. The demand for high-quality fuel can be achieved by the direct removal of sulfur present in liquid fuels. Fuel oil desulfurization employing various techniques will play an important role in the world's future clean and sustainable energy supply [1, 2]. Generation of sulfur-free oil has become an important part of worldwide energy policy to reduce sulfur concentration to less than 15 ppm given the stringent environmental regulations [3]. Sulfur compounds such as benzothiophene (BT), dibenzothiophene (DBT), and 4,6-dimethyldibenzothiophene (4,6-DMBT) found in fuel oil must be removed to keep environment safe. There are known desulfurization technologies namely; hydrodesulfurization, biodesulfurization, extractive, adsorptive, and oxidative desulfurization (ODS) that are seen as options to help in the removal of organosulfur compounds found in transportation fuel [4, 5]. Among the technologies, ODS has raised great interest due to mild operating conditions and the use of commercially available hydrogen peroxide. However, the reaction is slow and therefore a suitable catalyst is required to boost the reaction kinetics of the process [6].

Recently, metal-organic frameworks (MOFs) have received special attention especially on the catalytic field. Due to their structural diversity, chemical and thermal stability, high surface area, and excellent catalytic properties, Zr-based MOFs including MOF-808, NU-1000, and UiO-66 have been used in several technology applications especially in clean hydrogen generation, batteries, fuel cells, fuel storage, and waste decontamination [7]. Phosphotungstic acid (PTA) which is a Keggin-type polyoxometalates (POMs), possesses excellent redox properties with the potential for use in catalytic oxidation reactions. The Zr-MOF and the formation of its composite with POM to yield POM@MOF nanocomposites, are capable of demonstrating excellent catalytic properties and stability [8–10]. Impregnation and encapsulation methods have been reported as the most employed techniques for the incorporation of POMs into the pores of MOFs.

Herein, we reported the insertion of PTA into the mesoporous channels of MOF-808 by a simple impregnation method. The obtained product was acid treated to give PTA@MOF-808(Zr)-0.5H composite and was used for ODS reaction of refractory sulfur compounds under mild reaction conditions. In addition, the catalytic results showed that PTA@MOF-808(Zr)-0.5H can achieve up to 99.94% conversion of DBT at 50 °C within 30 min. The remarkable efficiency demonstrated is attributed to the addition of PTA which gives more catalytic active sites to Zr-MOF-808.

2 Materials and Method

All the chemicals and reagents used in the synthesis of MOF-808, PTA@MOF-808, and PTA@MOF-808(Zr)-0.5H are used as received from commercial suppliers. Zirconium (IV) chloride, 1,3,5-benzenetricarboxylic acid (BTC), N,N dimethylformamide (DMF), formic acid, hydrochloric acid, acetone, methanol, and PTA ($H_3PW_{12}O_{40}$) were purchased from Sigma Aldrich. For the catalytic oxidation reaction, BT, DBT, 4,6-DMBT, n-dodecane, acetonitrile (MeCN), and hydrogen peroxide (H_2O_2, 30 wt%) were purchased and used for experimental purposes only.

2.1 Synthesis and Preparation of MOF-808 and Composites

The synthesis of MOF-808(Zr) was carried out according to the previously published reports with some minor modifications [11]. In brief, a 0.1748 g of $ZrCl_4$ and 0.0525 g of BTC were mixed thoroughly with 16 mL of DMF/formic acid (1:1, v/v) and reacted both under continuous stirring and sonication each for 30 min. After sonication, the homogeneous mixture was then transferred into a 100 mL Teflon line autoclave and heated gradually to 120 °C for 24 h. After natural cooling, the product was washed vigorously with twice 30 mL each of DMF, acetone, and methanol to obtain pure MOF-808. The white powder was dried under a vacuum at 100 °C for 12 h. Furthermore, the as-synthesized MOF-808(Zr) was acid treated with 0.2 M HCl for 30 min to form MOF-808(Zr)-0.5H. The PTA@MOF-808(Zr)-0.5H was prepared by dissolving 0.045 g of PTA in 10 mL of distilled water, followed by the addition of 100 mg of MOF-808(Zr)-0.5H. The mixture was vigorously stirred for 24 h. The final product was isolated by centrifugation and washed with plenty of water. The product was oven dried at 100 °C for 6 h.

2.2 Characterization of Material

The physical and chemical properties of the obtained samples were investigated using powder X-ray diffraction (Xpert3) with a 2θ scanning range of 4–50° under Cu Kα radiation, and Fourier transform infrared (FTIR; Pekin Elmer model Frontier 01, USA) in the range of 4000–400 cm^{-1}. Field emission scanning electron microscopy (FESEM, Japan) was used for the observation of the morphology and size of the materials. The thermogravimetric analyzer was used to collect data (30–700 °C) at a heating rate of 10 °C per minute under nitrogen atmosphere to investigate the relationship between mass of the substance and temperature.

2.3 Catalytic ODS Reaction

The catalytic ODS experiment was performed in a three neck round bottom flask equipped with a reflux condenser and fitted with a magnetic stirrer at a given temperature. The sulfur-containing species injected in n-dodecane with an initial sulfur concentration of 500 ppm was removed by taking 5 mL of the solution, a catalyst was dispersed into the solution, followed by the addition of an appropriate amount of hydrogen peroxide. The mixture was stirred at 50 °C for a required period of time under vigorous stirring. After complete reaction, the mixture was cool down to room temperature and then transferred to a separating funnel followed by the addition of 5 ml of MeCN to extract the oxidized products. The oxidized sulfur compounds were each extracted and taken for GC-FID analysis. The sulfur removal efficiency was determined using the expression;

$$S(\%) = \frac{C_{s[\text{initial}]} - C_{s[\text{final}]}}{C_{s[\text{initial}]}} \times 100$$

where S is the percentage of desulfurization, $C_{s[\text{initial}]}$ and $C_{s[\text{final}]}$ are the initial and final concentration terms of sulfur compound at a specific duration.

3 Results and Discussion

3.1 Characterization of Samples

The FTIR and XRD patterns of MOF-808(Zr) and PTA@MOF-808(Zr)-0.5H are displayed in Fig. 1. The FTIR spectra (Fig. 1a) is recorded in the range of 4000 to 400 cm^{-1}. The absorption peaks in the range of 1621–1570 and 758 cm^{-1} are assigned to the asymmetric stretching vibration of carboxyl ground from BTC ensuring the coordination of BTC to the framework of the Zr-MOF. The absorption peaks at 660 and 717 cm^{-1} is an evidence of Zr–O bonds in Zr nodes for MOF-808 and has shown strong similarity with that of the composite. However, the Zr–O bonds nodes at around 717 cm^{-1} for the composite became weak following the incorporation of PTA and acid treatment. Moreover, the absorption peak located at 1576–1570 cm^{-1} represents the Zr–O–C bond between Zr_6 cluster and carboxyl group of BTC. The spectra and peak intensities of both the parent and composite materials are almost similar confirming that the structure of the MOF material is maintained after the incorporation of PTA and acid treatment with HCl. The XRD pattern of the prepared materials is depicted in Fig. 1b. The results confirm the formation of MOF-808(Zr) with high crystallinity and absence of impurity peak. A peak at around $2\theta = 4.3°$ was observed and assigned to the (111) crystal plane. In addition, the peaks appearing at $2\theta = 8.4°$, and 8.8° corresponds to (311) and (222) crystal planes of MOF-808(Zr), respectively.

The XRD pattern of PTA@MOF-808(Zr)-0.5H are in good agreement with that of MOF-808 with a slight change in peak intensity indicating the presence of guest particles. Overall, the diffraction peaks show good similarity and agreement with the previously reported literature [12].

Figure 2 shows the TGA curve of samples indicating weight loss in two steps. A weight loss of approximately 12 and 17% occurred between 220–330 °C and 430–570 °C, which corresponds to the loss of coordinated formate ion and the BTC linker, respectively [13]. Furthermore, a 50% weight loss was observed at a temperature greater than 600 °C. The images in Fig. 3 indicate a well-maintained morphology typical of an octahedral-like structure with a particle size of 300–400 nm. When PTA

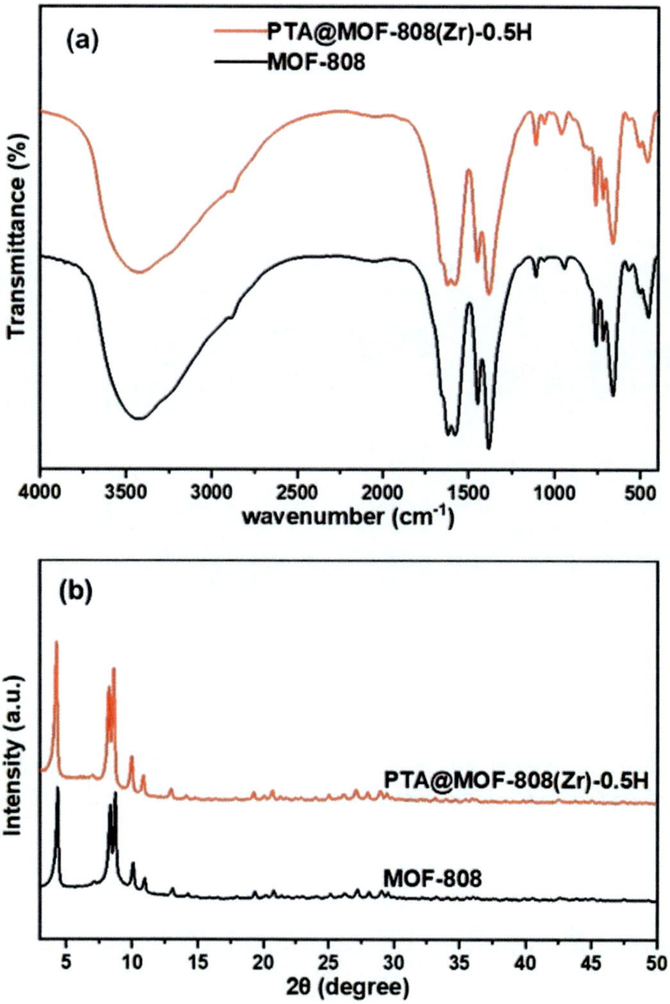

Fig. 1 **a** FTIR spectra, and **b** PXRD of MOF-808 and PTA@MOF-808(Zr)-0.5H

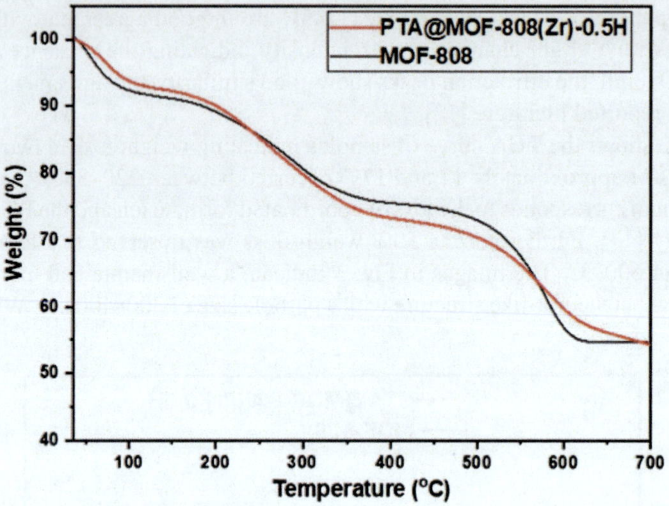

Fig. 2 TGA curves of MOF-808 and PTA@MOF-808(Zr)-0.5H

Fig. 3 **a** FESEM of MOF-808 and, **b** PTA@MOF-808(Zr)-0.5H with the uniform particle size of about 350 nm

was introduced and acid treated, an obvious change in morphology was observed as the particles aggregated to indicates composite formation. The structure was found to be in good agreement with the presented FTIR and PXRD results.

3.2 *Catalytic Result*

The catalytic performance for the ODS reaction of DBT was carried out at 50 °C using 30 mg of the catalyst. As shown in Fig. 4a, PTA and MOF-808(Zr) respectively

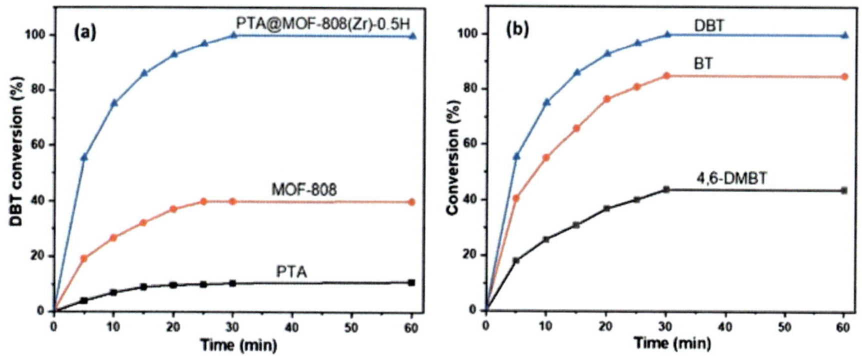

Fig. 4 Catalytic performance and comparison for the removal of DBT, BT, and 4,6-DMBT employing the prepared catalyst

yielded a 10 and 40% conversion of DBT in about 25 min of chemical reaction. Comparatively, PTA@MOF-808(Zr)-0.5H exhibits better ODS performance, and the conversion of 500 ppm DBT in n-dodecane reached 99.94% under the same reaction conditions. The excellent activity recorded was due to the promotion of sufficient active sites that makes contact with DBT and increase it conversion to sulfoxide. At the same time, the ODS performance of BT and 4,6-DMBT was tested and the removal efficiency follow the order 4,6-DMBT < BT < DBT (Fig. 4b). However, the remarkable efficiency demonstrated was due to the increased catalytic active sites and high chemical stability of the PTA@MOF-808(Zr)-0.5H compared to other POM@MOF catalysts used for ODS reaction.

4 Conclusion

In this research work, PTA@MOF-808-0.5H was successfully fabricated via an impregnation approach by incorporating PTA into the cavities of a Zr-MOF-808 followed by an acid treatment with 0.2 M HCl for 30 min. The obtained catalysts have been characterized to confirm their chemical structures using various analytical techniques. The oxidative desulfurization of n-dodecane containing BT, DBT, and 4,6-DMBT was evaluated using the as-synthesized catalyst employing H_2O_2 as an oxidant. The experimental results demonstrated that the catalysts exhibited sulfur removal efficiency greater than 99% and can promote the oxidation of DBT. In the past, only a few research were reported in utilizing an encapsulated PTA in zirconium-based MOFs for oxidative desulfurization studies. Interestingly, the catalytic efficiency of this new material shows remarkable improvement compared to the pristine POMs and MOFs. Since the generation of an ultra-clean fuel is significant to human survival and social developments, the produced catalysts provide inspiration for an environmentally friendly and green fuel technology. The catalytic system demonstrated high conversion efficiency toward fuel oil desulfurization and may have some potentials in hydrogen storage and generation.

Acknowledgements This research was fully funded and supported by YUTP research grant (FRG/1/2021/015LCC0-376). The Centre for Contaminant Control & Utilization (CenCoU) within the Institute of Contaminant Management, Universiti Teknologi PETRONAS is hereby gratefully acknowledged for all the supports. The first author is also grateful to the Universiti Teknologi PETRONAS for the PhD sponsorship through a graduate research assistantship.

References

1. L. Chen, Z.Y. Yuan, Design strategies of supported metal-based catalysts for efficient oxidative desulfurization of fuel. J. Ind. Eng. Chem. **108**, 1–14 (2022). https://doi.org/10.1016/j.jiec.2021.12.025
2. M.H. Suliman, C. Basheer, M.N. Siddiqui, A.A. Al-Arfaj, Biosynthesized silver nanoparticles decorated electro-membrane flow reactor an effective tool for the desulfurization of fuels. Arab. J. Sci. Eng. **47**, 543–550 (2022). https://doi.org/10.1007/s13369-021-06415-8
3. H. Li, Y. Li, L. Sun, S. Xun, W. Jiang, M. Zhang, W. Zhu, H. Li, H_2O_2 decomposition mechanism and its oxidative desulfurization activity on hexagonal boron nitride monolayer: a density functional theory study. J. Mol. Graph. Model. **84**, 166–173 (2018). https://doi.org/10.1016/j.jmgm.2018.07.002
4. A. Haruna, Z. Merican, A. Merican, S. Gani, Sulfur removal technologies from fuel oil for safe and sustainable environment. Fuel **329**, 125370 (2022). https://doi.org/10.1016/j.fuel.2022.125370
5. A. Haruna, Z.M.A. Merican, S.G. Musa, Recent advances in catalytic oxidative desulfurization of fuel oil—a review. J. Ind. Eng. Chem. **112**, 20–36 (2022). https://doi.org/10.1016/j.jiec.2022.05.023
6. B. Moeinifard, A. Najafi Chermahini, Mono lacunary phosphomolybdate supported on mesoporous graphitic carbon nitride: an eco-friendly and efficient catalyst for oxidative desulfurization of the model and real fuels, J. Environ. Chem. Eng. **9** (2021). https://doi.org/10.1016/j.jece.2021.105430
7. Y. Bai, Y. Dou, L.H. Xie, W. Rutledge, J.R. Li, H.C. Zhou, Zr-based metal-organic frameworks: design, synthesis, structure, and applications. Chem. Soc. Rev. **45**, 2327–2367 (2016). https://doi.org/10.1039/c5cs00837a
8. C.T. Buru, P. Li, B.L. Mehdi, A. Dohnalkova, A.E. Platero-Prats, N.D. Browning, K.W. Chapman, J.T. Hupp, O.K. Farha, Adsorption of a catalytically accessible polyoxometalate in a mesoporous channel-type metal-organic framework. Chem. Mater. **29**, 5174–5181 (2017). https://doi.org/10.1021/acs.chemmater.7b00750
9. Z. Zhang, X. Ma, X. Han, H. Cui, Y. Lu, S. Liu, Y. Liu, Straightforward construction of hollow polyoxometalate-based metal-organic framework via pseudo-homoepitaxial growth. Sci. China Chem. (2022). https://doi.org/10.1007/s11426-022-1295-3
10. S.G. Musa, Z.M. Aljunid Merican, A. Haruna, Investigation of isotherms and isosteric heat of adsorption for PW_{11}@HKUST-1 composite, J. Solid State Chem. **314** (2022), 123363. https://doi.org/10.1016/j.jssc.2022.123363
11. Z.Q. Li, J.C. Yang, K.W. Sui, N. Yin, Facile synthesis of metal-organic framework MOF-808 for arsenic removal. Mater. Lett. **160**, 412–414 (2015). https://doi.org/10.1016/j.matlet.2015.08.004
12. J.E. Efome, D. Rana, T. Matsuura, C.Q. Lan, Insight studies on metal-organic framework nanofibrous membrane adsorption and activation for heavy metal ions removal from aqueous

solution. ACS Appl. Mater. Interfaces. **10**, 18619–18629 (2018). https://doi.org/10.1021/acsami.8b01454

13. S. Karmakar, S. Barman, F.A. Rahimi, T.K. Maji, Covalent grafting of molecular photosensitizer and catalyst on MOF-808: effect of pore confinement toward visible light-driven CO_2 reduction in water. Energy Environ. sci. **14**, 2429–2440 (2021). https://doi.org/10.1039/d0ee03643a

Photoelectrocatalytic Properties of B@g-C_3N_4/PANI in CO_2 Reduction to Ethanol for Hydrogen Seasonal Storage

Mahmood Riyadh Atta, Maizatul Shima Shaharun, and M. D. Maksudur Rahman Khan

Abstract Recent interest is focused on hydrogen production, storage and utilisation. However, ethanol was found to be a very efficient liquid organic hydrogen carrier (LOHC) for seasonal energy storage. Therefore, producing ethanol from CO_2 is a good option, especially with photo-elecro-catalysis technology. In the present study, g-C_3N_4 modified with boron (B) and polyaniline (PANI) was investigated for photo-electro-chemical (PEC) CO_2 reduction to ethanol. The g-C_3N_4 was synthesised by the thermal decomposition of melamine and urea and the modification with boron and PANI was performed by the wet impregnation method. The catalysts were characterised by XRD, UV–visible spectrophotometer, chromatoampermetry, linear sweep voltametry (LSV), photoluminescence spectroscopy (PL), and Mott-Schottky analysis. A two-chamber PEC cell was used for PEC reduction of CO_2 where $NaHCO_3$ solution was used as an electrolyte. Visible light with a cut-off filter in the range of 420–670 nm was used. The band gap energy of g-C_3N_4, B@g-C_3N_4 and B@g-C_3N_4/PANI were found to be 2.73, 2.66 and 2.57 eV, respectively. As revealed by PL, the e^-/h^+ recombination rate was remarkably reduced due to the incorporation of B and PANI with g-C_3N_4. Additionally, compared to other catalysts, B@g-C_3N_4/PANI demonstrated higher photosensitivity as observed by chronoampermetry. In LSV, a significant increase in reducing current for all catalysts under CO_2 condition compared to the N_2 bubbled condition was observed, and the reducing current for B@g-C_3N_4/PANI under light on condition was drastically increased, suggesting the successful reduction of CO_2 under visible light.

M. R. Atta (✉) · M. S. Shaharun
Department of Fundamental and Applied Sciences, Universiti Teknologi PETRONAS, 32610 Seri Iskandar, Perak, Malaysia
e-mail: Mahmood_20000275@utp.edu.my

M. S. Shaharun
e-mail: maizats@utp.edu.my

M. D. Maksudur Rahman Khan
Faculty of Chemical and Natural Resources Engineering, Universiti Malaysia Pahang, Lebuhraya TunRazak, 26300 Gambang, Pahang, Malaysia
e-mail: mrkhancep@yahoo.com

Petroleum and Chemical Engineering, Faculty of Engineering, Universiti Teknologi Brunei, Bandar Seri Begawan, Brunei Darussalam

M. B. Othman et al. (eds.), *Proceedings of the 1st International Conference of New Energy*, Springer Proceedings in Energy, https://doi.org/10.1007/978-981-99-0859-2_14

Keywords CO_2 reduction · Ethanol production · Hydrogen carrier · Photoelectrochemical · g-C_3N_4 · Boron-doped · PANI polyaniline

1 Introduction

The increase of carbon dioxide (CO_2) emissions in the atmosphere causes global warming, a huge environmental issue with consequential environmental and societal changes. As a result, environmentalists have no choice but to decrease fossil fuel consumption to avoid and decrease CO_2 emissions [1]. Photoelectrocatalysis is the integration of electrocatalysis with photocatalysis by the usage of both energy sources of light and electricity potential, exploiting semiconductor electrodes instead of the usual conductor electrodes that are used in electrocatalysis [2].

However, the photoelectrocatalytic reduction of CO_2 decreases the consumption of electricity compared to the electrocatalytic reduction of CO_2 due to the introduction of solar energy in the photoelectrocatalysis. Photoelectrocatalysis, on the other hand, is more effective than photocatalysis because an external voltage can drive the separation of photo-generated electrons and holes, which is thought to be the most important step in making the process more effective [3].

Recently, researchers have been interested in studying the performance of metal-free catalysts in the photoelectrocatalysis of CO_2 reduction. However, the most important metal-free catalysts are the carbon-based catalysts. For instance, graphene, a sheet of carbon only one atom thick, has captured the imagination of scientists around the world. The remarkable thermal, mechanical, photosensitive and electrical capabilities of this material suggest numerous exciting future uses [4, 5]. A different type of carbon-based catalyst is graphene oxide (GO). GO is a novel material made through the acid exfoliation of graphite to form two-dimensional carbon sheets that are solution-dispersible. Graphitic carbon nitride (g-C_3N_4) is another carbon-based photocatalyst. Sun et al. argue that because of its significant potential at a low cost, g-C_3N_4 should be highlighted as a catalyst for CO_2 photocatalytic reduction. g-C_3N_4 can be developed and can achieve higher efficiency in CO_2 reduction by combining g-C_3N_4 with CO_2 adsorption/activation materials [6]. The g-C_3N_4 performance is high but not enough to be industrially applicable yet. It needs to be developed first and doped with co-catalyst.

Previous studies show that carbon-based catalysts are more effective in CO_2 reductions, such as Graphene (G), reduced grapheme oxide (rGO), and grapheme carbon nitride (g-C_3N_4). Nevertheless, the photocatalytic activity of g-C_3N_4 is higher than rGO and pure graphene, and it is low cost. It is nitrogen-rich and promising for CO_2 activation [7, 8]. The band structure of g-C_3N_4 is also good for photocatalytic reduction of CO_2 to different compounds with added value.

Furthermore, the researchers started to modify g-C_3N_4 with other elements and compounds to increase its efficiency as a catalyst. However, g-C_3N_4 does not have enough properties to be applied as a photoelectrocatalyst in its pure form because of its poor electroconductive [9]. As a result, the semiconductor polymer polyaniline

(PANI) is used to modify g-C_3N_4 with boron-doping into B@g-C_3N_4/PANI, which theoretically can be used as a photoelectrocatalyst for CO_2 reduction with low energy consumption and low cost [5].

2 Experimental

2.1 *Materials*

Cyanamide (CN_2H_2), ammonium persulfate ($(NH_4)_2S_2O_8$) (APS), aniline monomer, 5 wt% Nafion, Boric acid (H_3BO_3), isopropanol (C_3H_8O) and hydrochloric acid 1 M HCl were purchased from Merck, are all 99% is of analytical grade, absolute technical grade 99% and used as received. No further purification was required. Toray carbon paper and commercial 10% Pt/C were procured from Kuantan Sunny Scientific Collaboration Sdn. Bhd.

2.2 *Preparation of Boron-Doped-g-C_3N_4/PANI*

g-C_3N_4: To synthesise the metal-free g-C_3N_4 powders, 2.0 g of melamine was heated to 550 °C in an alumina combustion boat at a heating rate of 10 C/min for 4 h. Powder was made from the product after it was collected and processed [7].

Boron-doped g-C_3N_4: Co-condensation method was used to synthesise boron-doped g-C_3N_4. Typically, 25 mL of deionized water would be used to dissolve the boric acid powder before 5 g of melamine powder is added to create a homogenous solution that can be sonicated. After drying the mixture at 80 °C, it was placed in a covered alumina crucible and calcined at 500 °C for two hours, followed by two hours at 550 °C with a heating rate of 2.5 °C per minute. Unreacted contaminants were washed away by repeatedly washing the sample in isopropanol while it was cooled to room temperature [10].

Boron-doped-g-C_3N_4/PANI: To synthesise a boron-doped-g-C_3N_4/PANI composite photoelectrocatalyst, 1.0 g of boron-doped g-C_3N_4, 0.034 g ammonium persulfate $(NH_4)_2S_2O_8$ (APS) and 1.5 ml of 1 M HCl solution were mixed with 30 ml of distilled water in a reaction jar. To create a uniform suspension of B@g-C_3N_4 particles in the solution, it was agitated by hand for an hour in an ice water bath. Next, using a micro-injector, we added 0.01 ml of aniline monomer to the already cooled mixture. The resulting mixture spent 8 h chilling in an ice bath to react. Boron-doped g-C_3N_4/PANI composite powder was precipitated, then washed and filtered in ethanol and distilled water until the filtration solution turned clear. As a final step, vacuum dehydration was performed on the product [6].

2.3 Catalyst Characterisation

X-ray diffraction (XRD) was used to examine the catalysts and find out what caused the broadening of the diffraction peak [11]. Ultraviolet–visible (UV–vis) was used to measure the band gap of the catalysts. Photoluminescence spectrums were used to test the e^-/h^+ recombination process on the surface of each catalyst and the suitable wavelength for PEC reaction. The Mott-Schottky test was conducted with a platinum foil working electrode, a platinum foil counter electrode, an Ag/AgCl reference electrode, and a non-CO_2 bubbled 0.1 M $NaHCO_3$ solution (pH 6.8) as the electrolyte in an electrochemical analyser [12] (Autolab Compact PGSTAT 204, Netherland).

2.4 Photoelectrochemical Measurement

All studies involving the reduction of carbon dioxide by PECs were conducted in a single-chamber PEC cell fitted with a quartz window reactor. The prepared catalyst-electrode served as the working electrode, Ag/AgCl served as the reference electrode and platinum foil served as the counter electrode in $NaHCO_3$ aqueous solution for all of the PEC measurements. With a scan rate at 10 mV/s, 0.2 to −1.2 V were swept over in a linear sweep voltammetry (LSV) experiment. Then Chronoamperometry with an applied potential of −0.4 V and a 470 cut-off filter was utilised to evaluate the effect of light on the PEC reaction at each catalyst.

3 Result and Discussion

3.1 Characterization of B@g-C_3N_4/PANI

The XRD patterns of g-C_3N_4, B@g-C_3N_4 and B@g-C_3N_4/PANI are shown in Fig. 1. The black line indicates g-C_3N_4, which shows a sharp diffraction peak at 27°. The weak diffraction at 12.5° indicates the (002) and (100) planes (DB card number 9012236). In the XRD pattern of the B@g-C_3N_4 catalyst shown as the red line in Fig. 1, the 26.24° diffraction peak indicates the (002) crystal plane of the spinel B@g-C_3N_4 (DB-9011267). However, in the XRD pattern of the B@g-C_3N_4/PANI catalyst, weak diffraction peaks are shown at 26.24°, and 25.6° and 14.8° represent the phases of (002), (200) and (011) for B@g-C_3N_4/PANI. The (002) peak at 26.24° appeared in all catalyst XRD diffraction, indicating the presence of g-C_3N_4 [8, 13–16].

The Tauc plot and UV–visible spectra were performed on g-C_3N_4, B@g-C_3N_4 and B@g-C_3N_4/PANI to explore the optical properties of the photocatalyst as shown in Fig. 2a and b, respectively. Tauc plot was constructed using UV–vis data to determine the band gaps of the three catalysts. In Fig. 2a, it is observed that g-C_3N_4, B@g-C_3N_4 and B@g-C_3N_4/PANI have band gap energies of 2.73, 2.66 and 2.57 V, respectively.

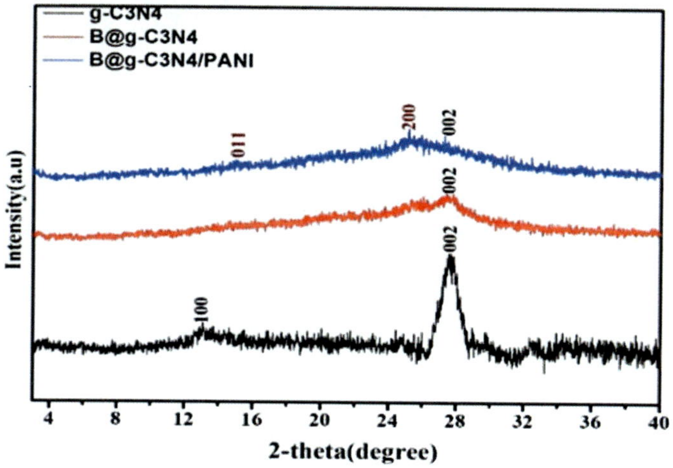

Fig. 1 XRD pattern of g-C_3N_4, B@g-C_3N_4 and B@g-C_3N_4/PANI

Meanwhile, Fig. 2b shows the UV–visible absorption spectra of g-C_3N_4, B@g-C_3N_4 and B-g-C_3N_4/PANI. The absorption wavelengths of the catalysts are in the range of 440–448 nm, which is considered strong evidence of visible light absorbance by all catalysts [10].

PL spectroscopy is to elucidate the effect of boron-doping and PANI incorporation of the g-C_3N_4 catalyst on the e^-/h^+ recombination process, as shown in Fig. 3a. The sharp emission peak at about 593 nm in the PL spectrum of g-C_3N_4 was produced by recombination of the photo-generated e^-/h^+ pair, and the intensity was remarkably lowered for the B@g-C_3N_4 sample, and lowered further for the B@g-C_3N_4/PANI sample. Due to the good interaction of boron and PANI with g-C_3N_4 nanoparticles, the low PL spectral intensity of B@g-C_3N_4/PANI indicates that the recombination rate of photoinduced charge carriers is low [17]. Mott-Schottky analysis was done

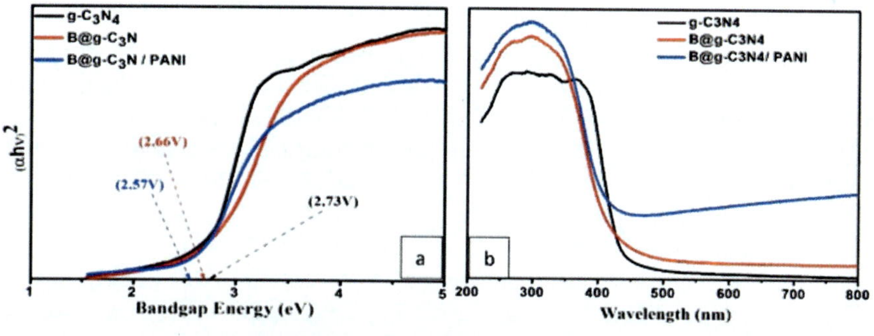

Fig. 2 **a** Tauc plot shows the band gaps of (g-C_3N_4, B@g-C_3N_4 and B@g-C_3N_4/PANI). **b** Uv–vis spectra of (g-C_3N_4, B@g-C_3N_4 and B-PANI-g-C_3N_4) wavelength range (200–800) nm

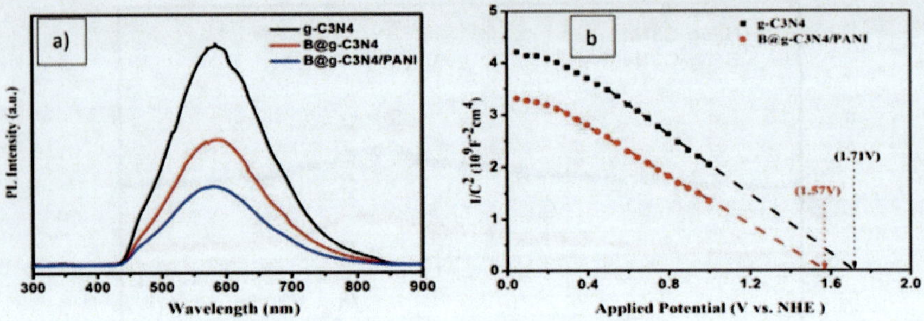

Fig. 3 **a** PL emission spectra (g-C_3N_4, B@g-C_3N_4 and B@g-C_3N_4/PANI). **b** Mott-Schottky plot of g-C_3N_4 and B@g-C_3N_4/PANI

at the same time for g-C_3N_4 and B@g-C_3N_4/PANI electrodes to find the flat band potential and confirm the type of semiconductor.

Figure 3b, shows the plot of the Mott-Schottky calculation. All catalysts were p-type semiconductors as they show negative slopes. From Fig. 3b, at 0 I/C^2 the applied potentials were observed at +1.71 and +1.57 V versus NHE for g-C_3N_4 and B-C_3N_4/PANI, respectively. However, the valance band (VB) position is located near the E_{fb} for *p*-type semiconductors [1, 18]. The E_{fb} values for g-C_3N_4 and B@g-C_3N_4/PANI were determined as +1.9075 and +1.76 V versus NHE, respectively. So, after the pH correction, the VB position for each catalyst was determined as 1.90 and 0.83 V for g-C_3N_4 and B@g-C_3N_4/PANI, respectively. Thus, the conduction bands (CB) of g-C_3N_4 and B@g-C_3N_4/PANI are identified based on band gab energies and VB position to be −0.83 and −1.74 V, respectively.

3.2 *Photoelectrochemical Performance*

In a 0.1 M $NaHCO_3$ solution purged with different gases (N_2 and CO_2), LSV further evaluated the effects of light illumination under dark and light conditions (470 nm cut-off filter) (Fig. 4a, b). In the N_2 bubbling system, the reduction current of the B@g-C_3N_4/PANI (−0.58 mA/cm^2) electrode under light conditions is higher than that caused by B@g-C_3N_4 (−0.48 mA/cm^2) which is further higher than that caused by g-C_3N_4 (−0.4 mA/cm^2). Moreover, in dark conditions, the reduction current was less in the three catalysts but they had the same trend. Meanwhile, in a CO_2 bubbling system, the reduction current of the B@g-C_3N_4/PANI (−0.49 mA/cm^2) electrode under light conditions is higher than that caused by B@g-C_3N_4 (−0.18 mA/cm^2) which is further higher than that caused by g-C_3N_4 (−0.09 mA/cm^2) which is in line with previous researches [19]. In dark conditions, the reduction current was lower in the three catalysts but had the same trends [4, 5].

For the sake of comparison, a significant current difference was observed during light irradiation compared to dark conditions. It is clear in Fig. 4a, that the light

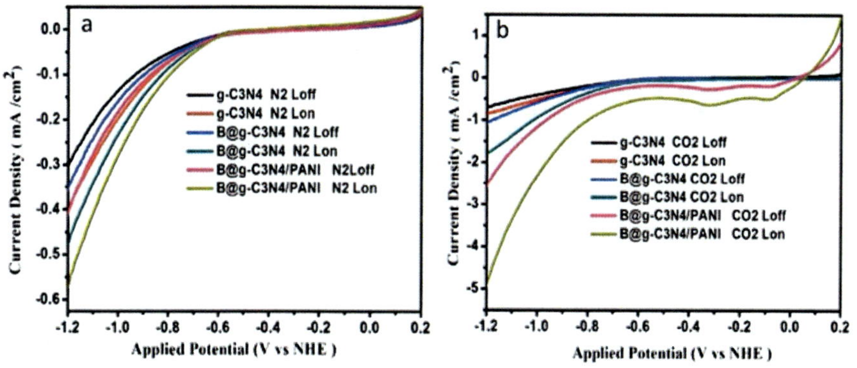

Fig. 4 **a** LSV analysis for N_2 reduction by (g-C_3N_4, B@g-C_3N_4 and B@g-C_3N_4/PANI) under light off/on. **b** LSV analysis for CO_2 reduction by same catalysts and conditions in **a**

response of the B@g-C_3N_4/PANI catalyst in N_2 reduction is (−0.4 to −0.58) mA/cm^2 which is higher than the other two catalysts. Figure 4b, depicts the B@g-C_3N_4/PANI current density light effect in CO_2 reduction, which increased from −0.25 to −0.5 mA/cm^2. But when CO_2 is reduced, the light response shows higher values of current density than when N_2 is reduced [20].

Using a 0.1 M $NaHCO_3$ solution, a potential of −0.40 V versus NHE was applied while bubbling CO_2 gas through the solution. The cathodic photocurrent quickly increases when light is shone on the electrodes, but then it drops off over time and stabilises at a low steady state value. Figure 5 shows that, within experimental error, there were slight variations in photocurrent generation during the first and second light on/off cycles [5].

Figure 5 shows the change in cathode photocurrent density over time for several optical switching cycles at an applied potential of −0.40 V over g-C_3N_4, B@g-C_3N_4

Fig. 5 Chronoamperometry of (g-C_3N_4, B@g-C_3N_4 and B-PANI-g-C_3N_4) with applied potential at −0.4 V and 470 cut-off filter

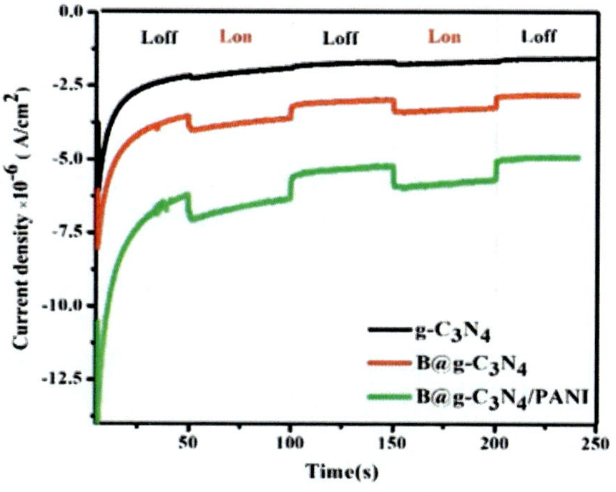

Table 1 Ethanol production rate (mmol/cm^2 h)

Electrode	Ethanol rate mmol cm^{-2} h^{-1}
g-C_3N_4	2.506158548
B@g-C_3N_4	3.004416294
B@g-C_3N_4/PANI	6.302833085

and B@g-C_3N_4/PANI catalysts. It was found that the photocathode photocurrent density rapidly increased under light irradiation. The current densities observed using B@g-C_3N_4/PANI are 27.9, and 11.17 μA using B@g-C_3N_4 and about 5.17 μA using the original catalyst g-C_3N_4, indicating that the incorporation of Boron as well as PANI improves the light-induced e^-/h^+ separation efficiency [6, 21].

4 Ethanol Production

Ethanol has many applications in energy and chemical reactions. However, one of the most interesting applications is using ethanol as a liquid organic hydrogen carrier (LOHC) for seasonal energy storage [22]. Therefore, ethanol production from the photo-electrocatalysis conversion of CO_2 over the prepared catalysts is a significant option. Table 1 shows the production rate of ethanol in mmole for each 1 cm^2 of electrode and 1 h of reaction. Data in Table 1 presents the impact of PANI enhancement on the photo-electro-catalysis properties of B@g-C_3N_4, as the ethanol produced by B@g-C_3N_4/PANI is doubled.

5 Conclusion

The graphitic carbon nitrate (g-C_3N_4) was successfully modified with boron-doping and Polyaniline (PANI) in a perfect combination of non-metallic substances to form a suitable photoelectrocatalyst mentioned as B@g-C_3N_4/PANI. Furthermore, the catalysts have been subjected to XRD, UV–vis, PL, and Mott-schottky analysis to identify the changes in the crystalline structure, electron–hole and band gap properties. Indeed, the XRD results show the perfect combination of boron, g-C_3N_4 and PANI, while UV–vis shows the reduction in band gab due to the modification from 2.73 eV for g-C_3N_4 to 2.53 eV for B@g-C_3N_4/PANI. PL results display the electron–hole recombination decrease in the modified catalyst B@g-C_3N_4/PANI, concluding that the photoinduced charge needed is reduced. In addition, from Mott-schottky analysis, the conduction band and valence band of B@g-C_3N_4/PANI were approximately 0.83 and −1.74 V, respectively. In LSV, a significant increase in reducing current for all catalysts under CO_2 condition compared to the N_2 bubbled condition was observed, and the reducing current for B@g-C_3N_4/PANI under light on condition was drastically increased, suggesting the successful reduction of CO_2 under visible light. In Chronoamperometry It was found that the photocathode photocurrent density increased rapidly under illumination. The photocurrent of B@g-C_3N_4/PANI is about 2.5 times that of B@g-C_3N_4 electrode and about 5.5 times higher than g-C_3N_4, indicating that light-induced e^-/h^+ separation efficiency can be improved by incorporation of boron and PANI. The production of ethanol shows the same trend where ethanol produced using B@g-C_3N_4/PANI is 2.15 times higher than B@g-C_3N_4, and 2.5 times higher than g-C_3N_4. More research is required to explore the effects of PANI loading and boron-doping on g-C_3N_4 and the mechanism phenomena for CO_2 reduction by PEC and study the application of ethanol as a liquid organic hydrogen carrier (LOHC) for seasonal energy storage.

References

1. Z. Chen, S. Zhang, Y. Liu, N.S. Alharbi, S.O. Rabah, S. Wang, X. Wang, Synthesis and fabrication of g-C(3)N(4)-based materials and their application in elimination of pollutants. Sci. Total Environ. **731**, 139054 (2020). https://doi.org/10.1016/j.scitotenv.2020.139054
2. IEA, World energy outlook special report (2015), p. 135
3. S. Xie, Q. Zhang, G. Liu, Y. Wang, Photocatalytic and photoelectrocatalytic reduction of CO2 using heterogeneous catalysts with controlled nanostructures. Chem. Commun. (Camb.) **52**(1), 35–59 (2016). https://doi.org/10.1039/c5cc07613g
4. M.M.R. Khan, M.R. Uddin, H. Abdullah, K.M.R. Karim, A. Yousuf, C.K. Cheng, H.R. Ong, Preparation and characterization of CuFe2O4/TiO2 photocatalyst for the conversion of CO2 into methanol under visible light. Int. J. Chem. Mol. Nucl. Mater. Metall. Eng. **10**(10), 1273–1280 (2016)
5. K.M.R. Karim, M. Tarek, S.M. Sarkar, R. Mouras, H.R. Ong, H. Abdullah, C.K. Cheng, M.M.R. Khan, Photoelectrocatalytic reduction of CO2 to methanol over CuFe2O4@PANI photocathode. Int. J. Hydrogen Energy **46**(48), 24709–24720 (2021). https://doi.org/10.1016/j.ijhydene.2020.02.195
6. S. Thaweesak, S. Wang, M. Lyu, M. Xiao, P. Peerakiatkhajohn, L. Wang, Boron-doped graphitic carbon nitride nanosheets for enhanced visible light photocatalytic water splitting. Dalton Trans. **46**(32), 10714–10720 (2017)
7. F. Liang, Y. Zhu, Enhancement of mineralization ability for phenol via synergetic effect of photoelectrocatalysis of g-C3N4 film. Appl. Catal. B **180**, 324–329 (2016). https://doi.org/10.1016/j.apcatb.2015.05.009
8. N. Sagara, S. Kamimura, T. Tsubota, T. Ohno, Photoelectrochemical CO2 reduction by a p-type boron-doped g-C3N4 electrode under visible light. Appl. Catal. B **192**, 193–198 (2016). https://doi.org/10.1016/j.apcatb.2016.03.055
9. W. Niu, K. Marcus, L. Zhou, Z. Li, L. Shi, K. Liang, Y. Yang, Enhancing electron transfer and electrocatalytic activity on crystalline carbon-conjugated g-C3N4. ACS Catal. **8**(3), 1926–1931 (2018). https://doi.org/10.1021/acscatal.8b00026
10. D. Liang, T. Jing, Y. Ma, J. Hao, G. Sun, M. Deng, Photocatalytic properties of g-C6N6/g-C3N4 heterostructure: a theoretical study. J. Phys. Chem. C **120**(42), 24023–24029 (2016). https://doi.org/10.1021/acs.jpcc.6b08699
11. H.G. Mohammed, T.M.B. Albarody, H.K.M. Al-Jothery, M. Mustapha, N.M. Sultan, A study of crystalline–texture and anisotropic properties of hexagonal BaFe12O19 sintered by in-situ magnetic-anisotropy spark plasma sintering (MASPS). J. Magn. Magn. Mater. **553**, 169268 (2022)
12. A.M. Al-Dhahebi, R. Jose, M. Mustapha, M.S.M. Saheed, Ultrasensitive aptasensor using electrospun MXene/polyvinylidene fluoride nanofiber composite for Ochratoxin A detection. Food Chem. **390**, 133105 (2022)

13. Y. Fang, X. Wang, Metal-free boron-containing heterogeneous catalysts. Angew. Chem. **56**(49), 15506–15518 (2017)
14. S. Zhang, L. Gao, D. Fan, X. Lv, Y. Li, Z. Yan, Synthesis of boron-doped g-C 3 N 4 with enhanced electro-catalytic activity and stability. Chem. Phys. Lett. **672**, 26–30 (2017). https://doi.org/10.1016/j.cplett.2017.01.046
15. Y. Mao, M. Wu, G. Li, P. Dai, X. Yu, Z. Bai, P. Chen, Photocatalytic degradation of methylene blue over boron-doped g-C3N4 together with nitrogen-vacancies under visible light irradiation. React. Kinet. Mech. Catal. **125**(2), 1179–1190 (2018). https://doi.org/10.1007/s11144-018-1414-0
16. X. Yu, P. Han, Z. Wei, L. Huang, Z. Gu, S. Peng, J. Ma, G. Zheng, Boron-doped graphene for electrocatalytic N2 reduction. Joule **2**(8), 1610–1622 (2018). https://doi.org/10.1016/j.joule.2018.06.007
17. J. Zou, Y. Yu, W. Yan, J. Meng, S. Zhang, J. Wang, A facile route to synthesize boron-doped g-C3N4 nanosheets with enhanced visible-light photocatalytic activity. J. Mater. Sci. **54**(9), 6867–6881 (2019). https://doi.org/10.1007/s10853-019-03384-0
18. T. Le Bahers, K. Takanabe, Combined theoretical and experimental characterizations of semiconductors for photoelectrocatalytic applications. J. Photochem. Photobiol. C **40**, 212–233 (2019). https://doi.org/10.1016/j.jphotochemrev.2019.01.001
19. M.R. Atta, M.S. Shaharun, M.M.R. Khan, B. Abdullah, A.F. Al-Mahmodi, N.D.M. Ridzuan, T.D. Munusamy, L.J. Wei, Enhancing the photo-electrocatalytic properties of g-C3N4 by boron doping and ZIF-8 hybridization. Inorg. Chem. Commun. 110235 (2022)
20. T.D. Munusamy, C.S. Yee, M.M.R. Khan, Construction of hybrid g-C3N4/CdO nanocomposite with improved photodegradation activity of RhB dye under visible light irradiation. Adv. Powder Technol. **31**(7), 2921–2931 (2020). https://doi.org/10.1016/j.apt.2020.05.017
21. N. Alizadeh, A. Salimi, R. Hallaj, A strategy for visual optical determination of glucose based on a smartphone device using fluorescent boron-doped carbon nanoparticles as a light-up probe. Mikrochim Acta **187**(1), 1–10 (2020)
22. B.L. Tran, S.I. Johnson, K.P. Brooks, S.T. Autrey, Ethanol as a liquid organic hydrogen carrier for seasonal microgrid application: catalysis, theory, and engineering feasibility. ACS Sustain. Chem. Eng. **9**(20), 7130–7138 (2021)

Progresses in Improving Mechanical Properties of Maraging Steel MS1 Through Laser Additive Manufacturing for Renewable Energy Application

Sarah Najm Al-Challabi, Pravin Mariappan, Thar Albarody, and Mohammad Shakir Nasif

Abstract Direct Metal Laser Sintering (DMLS), metal additive layer manufacturing, is a powder bed fusion process. It produces complex parts with a higher degree of manufacturing freedom compared to traditional subtractive manufacturing methods. The focus of this research is to analyze and compare the mechanical properties of Maraging Steel MS1 components utilizing DMLS with conventional methods. It is crucial to investigate the mechanical properties and microstructure of DMLS components. Such kind of data gives an idea of the surface topography and microstructure. Therefore, hardness measurements, atomic force microscopy, and mesoporous testing were performed to assess the microstructure and mechanical properties of Maraging Steel MS1 samples.

Keywords Additive manufacturing (AM) · Direct metal laser sintering (DMLS) · Mechanical properties · Microstructure properties · Maraging Steel MS1

1 Introduction

Additive Manufacturing (AM) or more commonly known as 3D printing, has gained popularity over the past years [1–3]. The attractive feature of this technology is to manufacture complex designs that are not possible by the conventional manufacturing methods such as milling and turning [4–6]. Being a class of powder bed fusion technology, it works on a principle of melting or sintering layers of material in layers

S. N. Al-Challabi (✉) · P. Mariappan · T. Albarody · M. S. Nasif
Mechanical Engineering Department, Universiti Teknologi PETRONAS, 31750 Tronoh, Malaysia
e-mail: sarah_21001298@utp.edu.my

P. Mariappan
e-mail: pravin_22585@utp.edu.my

T. Albarody
e-mail: dher.albarody@utp.edu.my

M. S. Nasif
e-mail: mohammad.nasif@utp.edu.my

M. B. Othman et al. (eds.), *Proceedings of the 1st International Conference of New Energy*, Springer Proceedings in Energy, https://doi.org/10.1007/978-981-99-0859-2_15

to form a complete 3D part. Powder bed fusion consists of selective laser melting (SLM), DMLS, selective laser sintering (SLS), and electron beam melting (EBM).

The 3D printing technology produces metal parts with complex geometry. However, the DMLS process is costly, and therefore, it is applicable when conventional methods are unable to produce complex components. The secondary concern of the DMLS process is variation in the mechanical properties compared to the same part using conventional methods [7].

Maraging steels get their name from martensitic and aging, which refers to the crystalline structure of the metal combined with the aging process to obtain the desired properties. MS1 is an iron and nickel alloy with minimal carbon content with attractive mechanical properties such as high strength, high fracture toughness, and dimensional stability [8]. The higher strength of MS1 is due to the formation of Ni_3(Mo, Ti) and Fe_2Mo intermetallic compounds precipitated during the aging process [2]. In addition, the low carbon content in Maraging steel will eliminate quench cracking, while the nickel is good for anti-corrosiveness. Maraging steel is very popular in the automobile and aerospace industries due to its high strength and high corrosion resistance.

Continuous research on the properties of Maraging steel, such as manufacturing process parameters, microstructure, microhardness, and tensile strength, have been investigated over the past decade. Kempen et al. examined the different aging conditions on the properties of Maraging steel and found that the ductility decreases with the increase in tensile strength due to age hardening. Previous studies [1–4] indicate that Maraging steel parts that undergo aging treatment have up to 99%-part density compared to the theoretical value.

Some DMLS samples have been found to have a higher ultimate tensile strength compared to conventional cast parts. The mechanical properties of as-built DMLS parts are very different from the conventional counterparts, but heat-treated DMLS components can manufacture products with properties similar to cast or milled components.

The current study aims to investigate the properties of DMLS parts and compare them to conventional metal parts. The actual mechanical properties of Maraging steel at the stated parameters are to be investigated. Also, it is necessary to differentiate the microstructure and microhardness of the Maraging steel at different planes as well as compare them to cast iron-nickel alloys. It will provide a broad understanding of the materials microstructure and mechanical properties for the identification of potential applications. Specifically, the current research is focused on investigating the mechanical properties of sintered Maraging Steel parts and analyzing the microstructure of sintered Maraging Steel samples.

2 Materials and Experiments

2.1 Preparation of Sample

An EOS M290 Direct Metal Laser Sintering (DMLS) machine was used for the fabrication of DMLS parts. Powder with a layer thickness of 40 μm (thickness can vary from 20 to 80 μm) and a Ytterbium fiber laser beam of 0.1 mm thickness was used. Atmospheric Nitrogen gas was used to bring oxygen level under 1.3% before the MS1 build initiation. Then the build plate, where the powder was intended to be sintered on, was heated above 40 °C. It is an important step to maintain the size and uniformness of the plate.

Parameters such as oxygen level and pressure at the optimum level are prerequisites to start the build manufacturing. 3D printed metal parts need to be sintered onto a build platform to avoid the denser sintered layer sinking in the powder bed.

Residual stresses can cause poor quality of DMLS parts, and therefore post-process heat is required to relieve stresses. In this study, the parts were stress relieved at 510 °C for 7 h and then air cooled. Next, the parts were separated from the base through electric discharge wire cutting machining (EDM). Five DMLS specimens were cut of dimension 7.5 × 12.5 × 3 cubic millimeters. The samples were subjected to mechanical testing and microstructure analysis. The chemical composition of Maraging Steel is given in Table 1.

2.2 Experiments

Following ASTM E384 standards [9], Vickers indenter with test forces of 1 kgf was applied for 15 s. The samples were polished through a series of abrasive paper assisted by lubricants. Etchant (1 g $CuCl_2$, 25 ml HNO_3, 50 ml HCL and 150 ml water) was applied to reveal the microstructure. Optical microscopy, pressure field emission scanning electron microscope (VPFESEM), Atomic Force Microscopy (AFM) and X-ray were performed to analyze the microstructure. Tensile testing was carried out to evaluate the strength of the samples. However, it was not feasible to obtain the 'dog bone' shape sample. Hence samples were prepared with dimensions closer to the standard. Mesoporous testing was conducted to find the pore size and pore concentration. The samples were kept in a vacuum flask filled with a fixed volume of nitrogen gas. The absorbed volume of nitrogen gas was an indicator to get the desired mesoporous testing results.

Table 1 Chemical composition in wt% of Maraging Steel MS1

Fe	Ni	Co	Mo	Ti	Al	Cr, Cu	C	Mn, Si	P, S
66.1	18	9	4.85	0.70	0.1	0.5	0.03	0.1	0.01

3 Results and Discussion

In this section, the mechanical properties such as hardness and tensile strength are discussed. Next, the microstructure profile of the Maraging Steel MS1 sample is analyzed using optical microscopy, FESEM, and AFM. Finally, mesoporous testing is discussed to evaluate the quality of the samples in terms of porosity.

3.1 Mechanical Properties

Results of the Vickers micro hardness testing is tabulated as shown in Table 2 and Fig. 1. The mean value of 3 samples were taken by taking 3 readings at 2 different points for each sample to reduce error in readings. Average hardness number of there 3 samples is obtained.

The graph shows the comparison between the MS1 theoretical and as-built hardness values. The hardness of as-built sample is significantly lesser (about 38.9%) when compared to the EOS Data Sheet of hardness values ranging from 50 to 57 HRC. This is due to the fact that the build parameters differ from the laboratory build parameters. Using the mean of the average hardness values to be 366.28 HV,

Table 2 Hardness testing results

Sample	Point	1	2	3	Average Vickers hardness
1	a	373.4	373.3	385.1	372.9
	b	375.2	369.6	360.7	
2	a	359.8	361.1	357.3	357.2
	b	355.4	349.8	359.6	
3	a	361.4	371.7	364.3	368.8
	b	368.7	367.3	379.3	

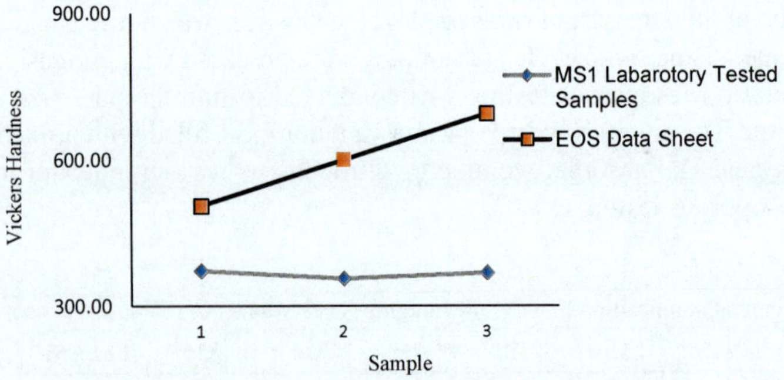

Fig. 1 Graph of hardness values for tested samples versus material data sheet

the equation discussed in article was deployed to obtain Rp0.2 which is the yield strength. The 0.2% Proof Strength then translates to the lower bound UTS estimate from using the correlation in Eq. 1 discussed in the methodology.

$$\mathrm{H} \approx 3 \times \mathrm{Rp0.2}$$

$$\mathrm{Rp0.2} \approx \frac{\mathrm{H}}{3} \tag{1}$$

$$\mathrm{Rp0.2} \approx \frac{366.28\mathrm{HV}}{3}\left(\frac{9.81\mathrm{H}}{\mathrm{HV}}\right)$$

$$\mathrm{Rp0.2} \approx 1197.7\,\mathrm{MPa}$$

$$\frac{\mathrm{UTS}}{\mathrm{Rp0.2}} = 1 + 2\left(\frac{150}{\mathrm{Rp0.2}}\right)^{2.5}$$

$$\mathrm{UTS} = 1211\,\mathrm{MPa}$$

Figure 1 shows the scatter of hardness and UTS results included in the graph of Mechanical Properties of DMLS versus Traditionally Produced Parts courtesy of MET, 2016. Maraging Steel exhibits similar properties as hardened steel which is in concurrent with the literature [5–7]. However, these numbers are way below the material data sheet number, since the sample produced was for industrial production and not research purposes (Fig. 2).

3.2 Atomic Force Microscopy (AFM) Profile

Results of AFM 3D phase diagram of as built and milled surfaces at 2 and 10 μm scale are shown in Fig. 3. The tool path for milling is evident in (d) and upon magnification shown in (c) we can see a high P–V values. Even though (b) appears to be smoother than (d) the results show that (b) is rougher. Higher magnification shows the granular structures of the as built surface at 2 μm scale as shown in Fig. 3.

Table 3 records the results of AFM parameters from the surfaces of as built and mechanical-milled Maraging Steel samples. From the table, there is a significant difference in the values of the AFM parameters when the milled surface was compared to the as built surface. The as built surface is rough due to the nature of DMLS process which was responsible for the powder splatter due to laser sintering. The as built sample has P–V value of almost 1.28 μm which suggested the rough surface. To reduce this, finer particles of powder and finer layer thicknesses can be

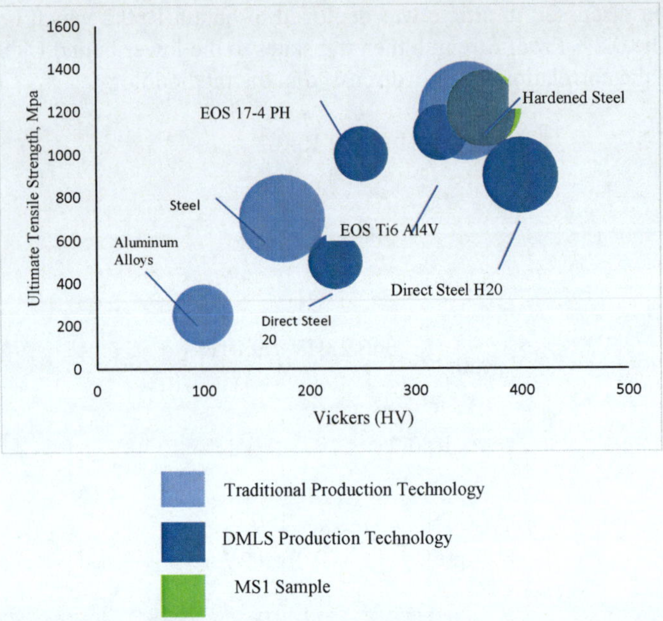

Fig. 2 Graph of UTS versus hardness for different metals

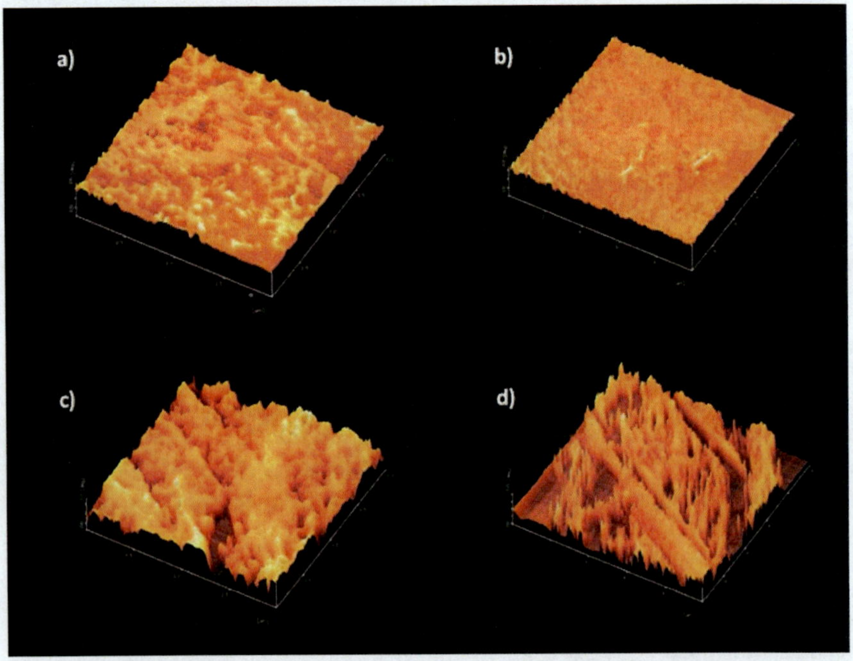

Fig. 3 AFM images of Maraging Steel as-built **a** 2 μm **b** 10 μm milled **c** 2 μm **d** 10 μm

Table 3 AFM parameters from the surfaces of as built and milled MS1 samples

Scan area (μm × μm)	Parameter	As built	Milled	%
10 × 10	Mean face roughness	151.3	52.05	66
	Peak-valley difference	1279.0	471.8	63
	Square mean face roughness	186.4	65.7	65
	Mean roughness at N-points	798.6, N = 6	182.6, N = 10	77
2 × 2	Mean face roughness	61.7	18.42	70
	Peak-valley difference	364.6	159.8	56
	Square mean face roughness	74.5	23.5	68
	Mean roughness at N-points	364.6, N = 2	69.3, N = 10	81

used but this will increase production cost. Hence, postproduction processes such as milling and mechanical polishing are better.

3.3 *Microscopic Profile*

Optical microscopy assessment was performed on the Maraging steel sample. Upon analysis it was found that there were two defects, namely unmelted particles and cavities. Spherical cavities ranging from 12 to 25 μm can be observed from Fig. 4a, b. These said cavities reduced the overall density of the part and are the initiators to cracks and fatigue. These cavities were suggested to be due to gas pockets that are formed in the powder feed layers. Another reason is the fluctuations in power of the laser that affected the proper fusing rate of the metal. The effect of the early onset of precipitate coarsening as suggested by Mooney et al., showing that the aging process made the alloy weak due to the softening behaviour [8]. Moreover, low-temperature aging encourages the nucleation and growth of the precipitate and allows a populous dispersion of small precipitates.

3.4 *Mesoporous Testing*

Mesoporous testing was conducted using the Micrometrics ASAP 2020 at a room temperature of 24 °C and a humidity ranging from 45 to 70% rH. The specimen was cleaned and prepared for testing. This procedure uses the principle of adsorption and desorption, of nitrogen, to measure pore volume and pore size. The pore surface area and volume are tabulated in Tables 5 and 6 respectively. The total cumulative surface area and cumulative volume of pore was determined by Barrett–Joyner–Halenda (BJH) model, while the sorption level pore width was determined by Brunauer–Emmett–Teller (BET) model. The adsorption average pore diameter of 4.4981 nm

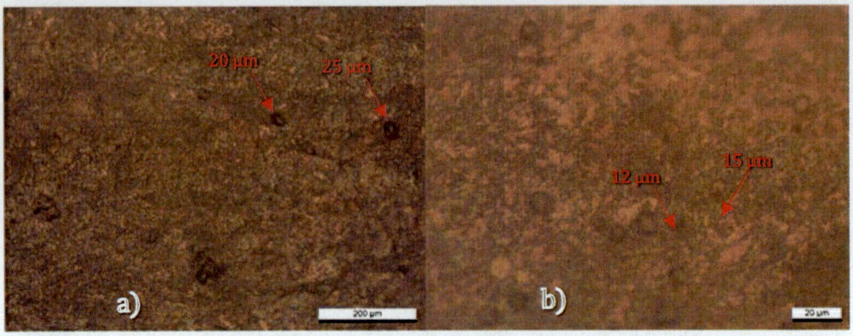

Fig. 4 Examples of defects found on the surface of the sample. **a** 10× magnification; **b** 50× magnification

Table 4 Pore size characteristics

Pore size	
Adsorption average pore width (4 V/A by BET)	0.98868 nm
Desorption average pore width (4 V/A by BET)	0.97468 nm
BJH adsorption average pore diameter (4 V/A)	4.4981 nm
BJH desorption average pore diameter (4 V/A)	3.4959 nm

was observed, while the desorption average pore diameter was 3.4959 nm. Hence, the average pore diameter was about 3.997 nm while the average pore width was 0.98168 nm. As a result, we can confirm the presence of meso-pores in the sample of Maraging Steel. As shown in Tables 4, 5 and 6.

Figure 5 shows the N_2 sorption results of mesoporous Maraging Steel sample from mesoporous testing. Ideally the desorption curve should have intersected with the adsorption curve at the lower relative pressure. However, this did not happen as the relative pressure at the end of desorption was 0.15 which was higher when compared

Table 5 Pore volume characteristics

Pore volume	
1-point adsorption total pore volume of pores below 139.3732 nm, diameter at P/Po = 0.985916613	0.64
1-point desorption total pore volume of pores less than 65.5153 nm, diameter at P/Po = 0.969563419	0.000063 cm^3/g
t-Plot micropore volume	−0.000094 cm^3/g
BJH adsorption cumm. Pore volume between 1.7000 nm and 300.0000 nm diameter	0.000191 cm^3/g
BJH desorption cumm. Pore volume between 1.7000 nm and 300.0000 nm diameter	0.000046 cm^3/g

Table 6 Pore surface area characteristics

Surface area	
Single point surface area at P/Po = 0.249702296	0.1376 m^2/g
BET surface area	0.2570 m^2/g
Langmuir surface area	1.0079 m^2/g
t-Plot external surface area	0.3444 m^2/g
BJH adsorption cumulative surface area of pores between 1.7000 nm and 300.0000 nm diameter	0.170 m^2/g
BJH desorption cumulative surface area of pores between 1.7000 nm and 300.0000 nm diameter	0.0525 m^2/g

with the value of 0.01 at the beginning of adsorption. A possible explanation is that the N_2 adsorbed is trapped within the pores of the Maraging Steel during desorption.

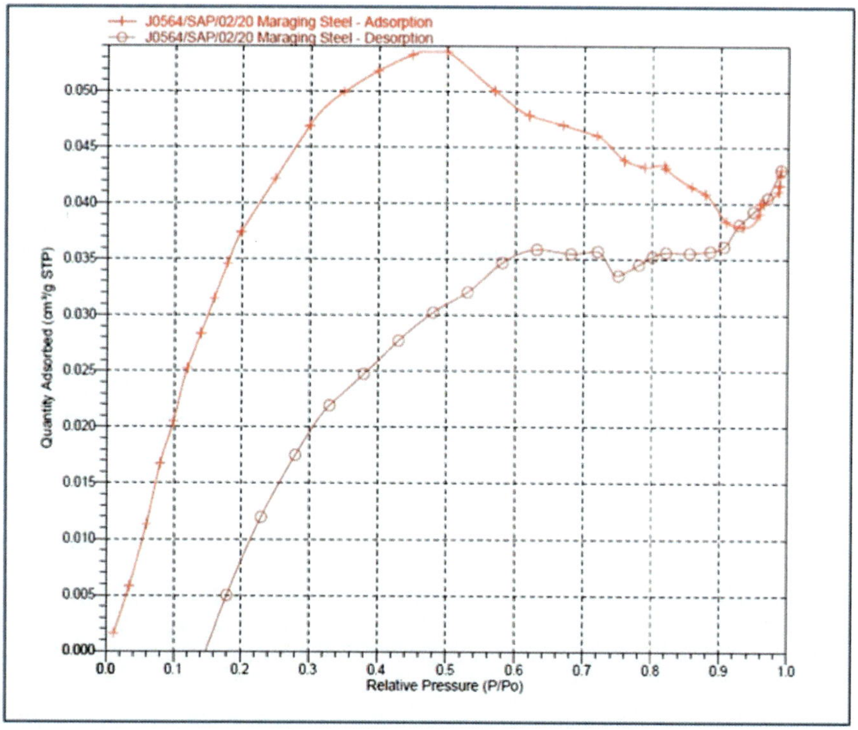

Fig. 5 N_2 adsorption–desorption isotherm linear plot of Maraging Steel

4 Conclusion

Based on the previous discussion, the following conclusions are drawn

- DMLS successfully developed components Maraging steel.
- DMLS exhibits the same hardness and tensile strength of traditionally manufactured hardened steel which are about 366.28 HV and 1211 MPa respectively.
- Based on microstructure analysis, cavities ranging from 12 to 25 μm were found which are assumed to be a side effect of layered sintering.
- Mean surface roughness 10 by 10-μm was achieved that is better than traditionally manufactured metals. However, the roughness still limited by the application of the parts produced.
- The dimension of pore (diameter ~3.997 nm and depth ~0.98168 nm) measured through mesoporous testing showed the existence of mesoporous in the DMLS-manufactured Maraging Steel

Numerical modeling based on advanced mechanical and material characterization is needed for finding optimum process parameters for manufacturing Maraging steel to identify more potential application especially in the field of renewable energy.

References

1. A. Cucchi et al., Clinical and volumetric outcomes after vertical ridge augmentation using computer-aided-design/computer-aided manufacturing (CAD/CAM) customized titanium meshes: a pilot study. BMC Oral Health **20**(1) (2020). https://doi.org/10.1186/s12903-020-01205-4
2. M. Ghasri-Khouzani et al., Comparing microstructure and hardness of direct metal laser sintered AlSi10Mg alloy between different planes. J. Manuf. Process. **37**, 274–280 (2019). https://doi.org/10.1016/j.jmapro.2018.12.005
3. B. Stieberova, M. Broumova, M. Matousek, M. Zilka, Life cycle assessment of metal products produced by additive manufacturing: a metal mold case study. ACS Sustain. Chem. Eng. **10**(16), 5163–5174 (2022). https://doi.org/10.1021/acssuschemeng.1c08445
4. A. Gratton, Comparison of mechanical, metallurgical properties of 17–4PH stainless steel between direct metal laser sintering (DMLS) and traditional manufacturing methods. Proc. Natl. Conf. Undergrad. Res. **2012**(July), 423–431 (2012)
5. H. Jaber, J. Kónya, K. Kulcsár, T. Kovács, Effects of annealing and solution treatments on the microstructure and mechanical properties of Ti6Al4V manufactured by selective laser melting. Materials (Basel) **15**(5) (2022). https://doi.org/10.3390/ma15051978
6. B. Vrancken, L. Thijs, J.P. Kruth, J. Van Humbeeck, Heat treatment of Ti6Al4V produced by selective laser melting: microstructure and mechanical properties. J. Alloys Compd. **541**, 177–185 (2012). https://doi.org/10.1016/j.jallcom.2012.07.022
7. B. Mooney, K.I. Kourousis, R. Raghavendra, D. Agius, Process phenomena influencing the tensile and anisotropic characteristics of additively manufactured maraging steel. Mater. Sci. Eng. A **745**, 115–125 (2019). https://doi.org/10.1016/j.msea.2018.12.070

8. X. Xu, S. Ganguly, J. Ding, S. Guo, S. Williams, F. Martina, Microstructural evolution and mechanical properties of maraging steel produced by wire + arc additive manufacture process. Mater. Charact. **143**(October 2017), 152–162 (2018). https://doi.org/10.1016/j.matchar.2017.12.002
9. ASTM, *ASTM E384-09, Standard Test Method for Microindentation Hardness of Materials* (ASTM International, West Conshohocken, PA, 2015), www.astm.org (2009)

Non-noble Metal Nanoparticles Formed in Interlayer of Layered Double Hydroxide for Hydrogen Production via Sodium Borohydride Hydrolysis Reaction

Hitoshi Inokawa, Aishah Mahpudz, Ryuichi Tomoshige, and Katsuki Kusakabe

Abstract Catalytic performance of non-noble metals has been required to be improved for their industrial use. In this study, synthesis techniques of fine nanoparticles of various transition metals in a layered double hydroxide (LDH) have been investigated for catalytic appreciation toward sodium borohydride hydrolysis reaction producing hydrogen. Citrate complexes of transition metals, which were iron, cobalt, nickel, or copper, were intercalated into interlayer spaces of LDHs, and were reduced by sodium borohydride in order to form their metallic nanoparticles. X-ray diffraction suggested that the citrate complexes were successfully intercalated into the interlayer of the LDHs by a conventional ion exchange treatment, and infrared spectroscopy showed that the citrate ligand was decomposed via the reduction process. Transmission electron microscope observation revealed that their metallic fine particles of several nanometers in diameter were obtained on the LDHs after the reduction although large particles of 100–200 nm in diameter were also formed from the citrates of iron and copper. Nitrogen adsorption/desorption isotherms showed a significant increase in surface areas and pore volumes by forming the metallic nanoparticles, suggesting that the nanoparticles were formed in the interlayer of LDHs. As a result of catalytic characterization, the cobalt nanoparticles showed the highest hydrogen generation rate for the hydrolysis reaction of sodium borohydride. Analysis of Arrhenius plots revealed that the cobalt nanoparticle had the lowest activation energy of the four. Therefore, cobalt nanoparticle supported in interlayer of LDHs is a promising catalyst for hydrogen generation via the sodium borohydride hydrolysis.

H. Inokawa (✉) · A. Mahpudz · R. Tomoshige · K. Kusakabe
Division of Applied Chemistry, Gradated School of Engineering, Sojo University, 4-22-1, Ikeda, Nishi-ku, Kumamoto 860-0082, Japan
e-mail: inokawa@nano.sojo-u.ac.jp

A. Mahpudz
e-mail: aishah.mahpudz@utp.edu.my

R. Tomoshige
e-mail: tomosige@nano.sojo-u.ac.jp

K. Kusakabe
e-mail: kusakabe@nano.sojo-u.ac.jp

M. B. Othman et al. (eds.), *Proceedings of the 1st International Conference of New Energy*, Springer Proceedings in Energy, https://doi.org/10.1007/978-981-99-0859-2_16

Keywords Cobalt · Iron · Nickel · Copper · Catalysis · Porous composite

1 Introduction

Hydrolysis reaction of sodium borohydride ($NaBH_4$) has attracted attention because of its high hydrogen content (10.8 mass%) and other advantages as a hydrogen generator for fuel cells described as follows [1]. The hydrogen production via the hydrolysis reaction can carry out at room temperature whereas hydrogen production from ammonia and methylcylohexane requires heating. The $NaBH_4$ hydrolysis generates pure H_2 without carbon monoxide or ammonia, which cause significant damages on fuel cells. The generated hydrogen can be directly supplied to fuel cells without any purification or filtration. The aqueous solution of $NaBH_4$ can be stabilized by the addition of NaOH, which realizes that $NaBH_4$ aqueous solution is stably stored as a liquid fuel. Hydrogen generation can be easily controlled by catalysts.

Although noble metals such as platinum and ruthenium show superior performance [1–3], their industrial application is limited because of the limitation of their source. Improvement of non-noble metal catalysts have been on a demand. Generally, the size of the catalyst particle is one of the important parameters affecting its activity because of the increase in surface area, surface energy, and so on.

We focused on layered double hydroxides (LDHs) as a catalyst support of the metal nanoparticles because of their interaction with negatively charged materials, such as BH_4^-. Although supporting materials are generally added into aqueous solutions of metal salts, e.g. nitrate or chloride, by conventional impregnation methods, LDHs has difficulty as the catalyst support because they can be dissolved in the metal salts' solutions due to a low pH condition. In addition, cationic precursors, such as Co^{2+} or Ni^{2+}, cannot be directly intercalated into the interlayer of LDHs because the interlayer is anionic space. We previously succeeded to synthesize cobalt nanoparticles in the interlayer of LDHs via intercalation of cobalt citrate anionic complex and reduction using $NaBH_4$ [4]. We also reported hydrogen generation properties of the cobalt nanoparticles with varied particle sizes [5, 6].

In this study, a variety of elements, which compose nanoparticles in the interlayer of LDHs, has been investigated. In detail, metallic nanoparticles of iron, cobalt, nickel, or copper were synthesized from their citrate complexes in interlayer of LDH composed of magnesium and aluminum (MgAl-LDH) as shown in Scheme 1, and their catalytic properties were investigated.

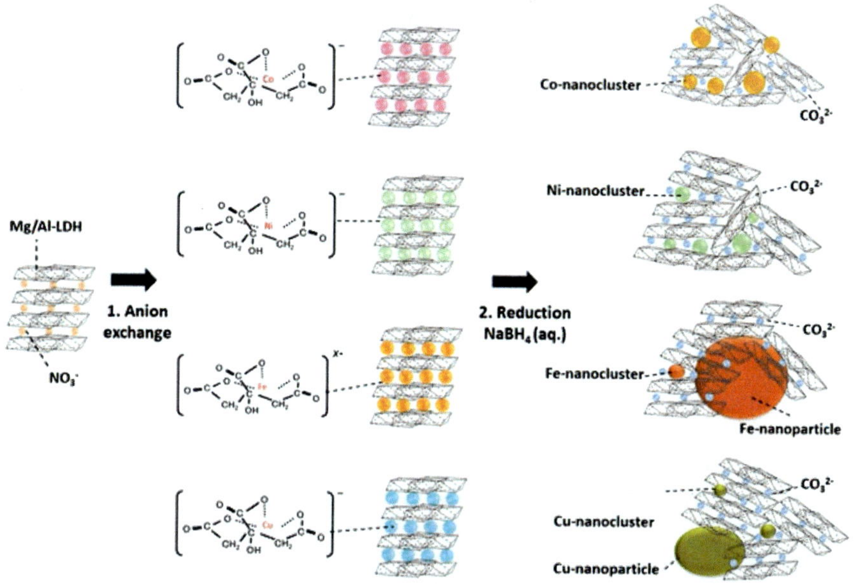

Scheme 1 Synthesis procedure of metal nanoparticles in interlayer of LDHs

2 Experimental

2.1 Materials

Sodium hydroxide (NaOH, 97.0%), Sodium nitrate ($NaNO_3$, 99.0%), magnesium nitrate hexahydrate ($Mg(NO_3)_2{\cdot}6H_2O$, 99.0%), aluminum nitrate nonahydrate ($Al(NO_3)_3{\cdot}9H_2O$, 98.0%), cobalt (II) nitrate hexahydrate ($Co(NO_3)_2{\cdot}6H_2O$, 98.0%), copper (II) nitrate trihydrate ($Cu(NO_3)_2{\cdot}3H_2O$, 77–80%), iron (III) nitrate nonahydrate ($Fe(NO_3)_3{\cdot}9H_2O$, 99%), nickel (II) nitrate hexahydrate ($Ni(NO_3)_2{\cdot}6H_2O$, 99%) and citric acid ($C_6H_8O_7{\cdot}H_2O$, 99.5%) were purchased from Kanto Chemical Co. Sodium borohydride ($NaBH_4$, 98%) were purchased from Acros Organics. All reagents were used without further purification. For catalyst synthesis, deionized water was degassed prior to use. This is to prevent carbonate (CO_3^{2-}) contamination in the resulting sample.

2.2 Catalyst Preparation

MgAl-LDH with a Mg/Al molar ratio of 2.0 was synthesized via a typical co-precipitation method in $NaNO_3$ aqueous solution in order to intercalate NO_3^- in the interlayer. In general, this procedure involves a dropwise addition of a mixed salt solution of $Mg(NO_3)_2{\cdot}6H_2O$ and $Al(NO_3)_3{\cdot}9H_2O$ (100 ml, 0.5 M) with NaOH solution (100 ml, 2 M), simultaneously into a reaction vessel containing $NaNO_3$ (250 ml, 1 M). Throughout the synthesis, the mixture was stirred vigorously and was kept at a constant temperature of 55 °C while the pH was maintained at 10 by varying the flow rates of both solutions. After filtration, rinsing with deionized and degassed water, and drying at 100 °C, a white powder was obtained, which is labeled as (NO_3^-)-LDH.

For anion exchange, metal citrate complex anion (M-citrate)$^-$ solution was first prepared. For this, 4 mM of metal nitrate (either Co, Fe, Ni or Cu) and equimolar ratio of citric acid were mixed with 300 ml of deionized water in a flask and placed on a water bath maintained at 50 °C. Then, pH of the solution was increased to 9.5 by using 0.1 M of NaOH prior to adding 1.0 g of dried (NO_3^-)-LDH. The resulting suspension was stirred for 12 h under strict nitrogen environment. Following that, it was filtered and rinsed with deionized water. For the purpose of characterization, some of the solid residue was collected and then dried at 60 °C overnight. These samples are referred as (M-citrate)- LDH where M is either Co, Fe, Ni, or Cu. The other half of the remaining solid residue was immediately reduced according to the following procedure.

To obtain the respective metal nanoparticles supported on LDH, the wet (M-citrate)-LDH was chemically reduced at 50 °C in $NaBH_4$ solution, which was prepared by dissolving 5 g of $NaBH_4$ into 50 ml of water. After stirring for 2 h with nitrogen flow, the suspension was filtered to collect the solid residue, thoroughly rinsed with deionized water, and then dried at 60 °C for 12 h. These samples are referred to as M-LDH where M is either Co, Fe, Ni, or Cu.

2.3 *Structural Characterization*

X-ray diffraction (XRD, Rigaku Smartlab) analysis was performed with utilizing Cu Kα irradiation. Nitrogen physisorption isotherms were obtained at –195.8 °C using surface area and porosity analyzer (Micromeritis TriStar II). Structure of anions in interlayer of LDH was characterized by Fourier transform infrared (FT-IR, Perkin Elmer Spectrum 100). The mass fraction of metal nanoparticle was determined from Inductively coupled plasma-optical emission spectrometry (ICP-OES, ICAP7000 Thermo Scientific). For ICP-OES analysis, the analysis solution was prepared by dissolving 10 mg of the sample in 100 ml of 0.1 M nitric acid and digested overnight. The shape and size of the metal nanoparticles were observed using transmission electron microscope and scanning transmission electron microscope (TEM/STEM, FEI Titan Themis 200).

2.4 Evaluation of Hydrogen Generation Property

$NaBH_4$ hydrolysis was carried out at 25 °C to evaluate the catalytic activity of the M-LDH catalysts. The volume of the generated hydrogen was measured using the water displacement method. For each run, 0.05 g of catalyst was added into 10 g of an aqueous solution containing 5 mass% of $NaBH_4$ and 5 mass% NaOH. The accumulated volume of hydrogen generated over time was equivalent to the volume of water displaced into a mass balance. The weight of displaced water was recorded over time and the hydrogen generation rate (HGR) was calculated according to Eq. 1. In order to evaluate activation energy, HGR was evaluated at various temperature and Arrhenius plots were analyzed.

$$\mathrm{HGR} = \frac{slope\ of\ hydrogen\ generation}{weight\ of\ metal(M)} \quad (1)$$

3 Results and Discussion

3.1 Structural Characterization

XRD patterns of samples at each preparation step are shown in Fig. 1. In all cases of (M-citrate)-LDHs, peaks assigned to (0 0 3) and (0 0 6) were shifted to a lower angle compared to those of (NO_3^-)-LDH. This suggests an expansion of interlayer space of LDHs, meaning that M-citrate complexes were successfully intercalated into the interlayer of LDHs. After the reduction, XRD patterns obtained from M-LDHs were same as the reference pattern of hydrotalcite, indicating that anions in interlayer were transformed from nitrate to carbonate. Any peak assigned to metallic Fe, Co, or Ni was not observed on the XRD patterns of M-LDHs, whereas a small peak of metallic Cu appeared at 43° on Cu-LDH. This suggests very low crystalinity of Fe, Co, and Ni species.

IR spectra at each preparation step shown in Fig. 2 suggest that citrate was taken into the LDHs after the ion exchange process and decomposed to carbonate anion via the reduction process according to the appearance and disappearance of a peak assigned to COO^-.

TEM/STEM images and size distribution of metal nanoparticles shown in Fig. 3 revealed that fine metal nanoparticles of 2–5 nm in diameter were obtained over all M-LDH, where M is Fe, Co, Ni, or Cu. In case of Fe-LDH and Cu-LDH, huge particles of 100–200 nm in diameter were also formed as shown in Fig. 3a, e.

Both BET surface areas and pore volumes evaluated from the nitrogen adsorption/desorption isotherms drastically increased after forming metal nanoparticles compared to those of (NO_3^-)-LDH, as shown in Table 1. In addition, the isotherms

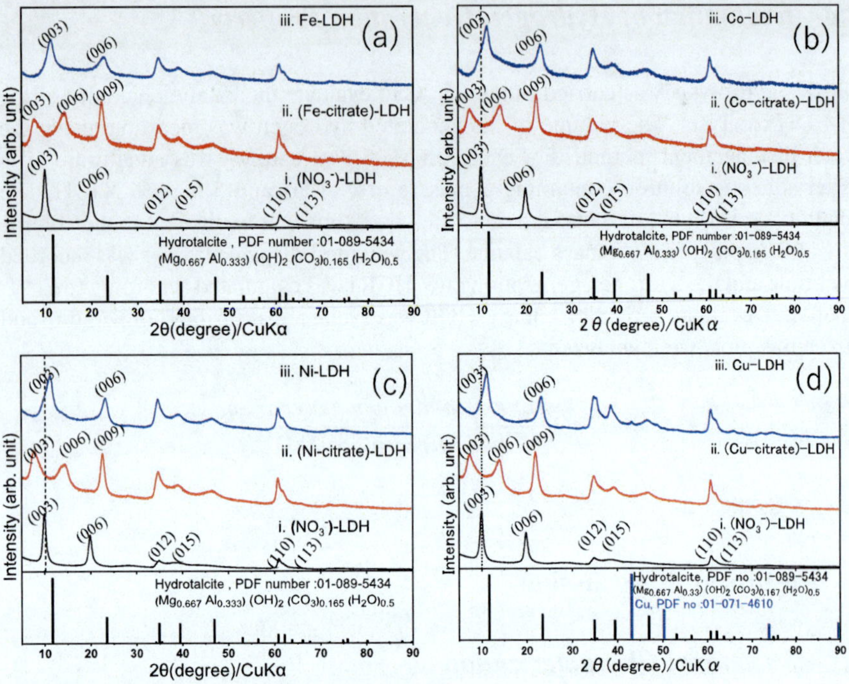

Fig. 1 XRD patterns of LDHs containing **a** Fe, **b** Co, **c** Ni, and **d** Cu species with a reference pattern of hydrotalcite. In each graph, labels of i, ii, and iii mean (NO_3^-)-LDH, (M-citrate)-LDH, and M-LDH, respectively

of all M-LDHs showed H3 type hysteresis. Therefore, it was indicated that M-LDHs had porous structure with slit type pores.

As a summary of the structural characterization, metal nanoparticles of Fe, Co, Ni, and Cu were formed with their fine size (<5 nm) in the interlayer of LDHs, which is consistent with our concept, whereas Fe and Cu also formed large particles, as shown in Scheme 1. Reduction condition should be optimized for Fe-LDH and Cu-LDH.

3.2 Hydrogen Production Property

Hydrogen generation properties of non-catalytic or catalytic $NaBH_4$ hydrolysis are shown in Fig. 4. No hydrogen was generated without catalyst or with the addition of (NO_3^-)-LDH, showing that MgAl-LDH had no catalytic activity. Hydrogen generation rates (HGRs) of M-LDHs were in order of Co-LDH (3.725 L min^{-1} $g_{Co}{}^{-1}$) > Fe-LDH (0.145 L min^{-1} $g_{Fe}{}^{-1}$) > Ni-LDH (0.007 L min^{-1} $g_{Ni}{}^{-1}$) > Cu-LDH (0.003 L min^{-1} $g_{Cu}{}^{-1}$). Activation energy evaluated from Arrhenius plots of M-LDHs was in order of Co-LDH (51.6 kJ mol^{-1}) < Fe-LDH (54.8 kJ mol^{-1}) < Ni-LDH (62.3 kJ

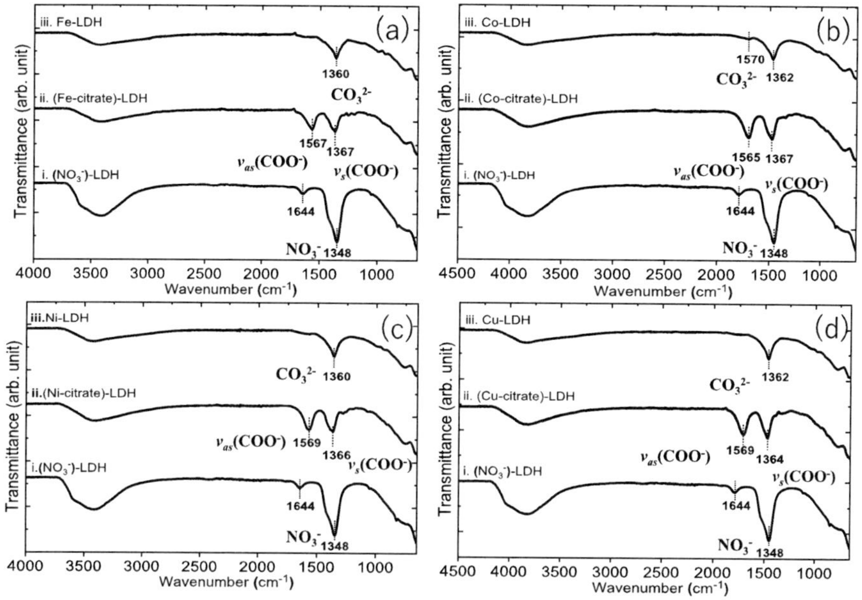

Fig. 2 IR spectra of LDHs containing **a** Fe, **b** Co, **c** Ni, and **d** Cu species. In each graph, labels of i, ii, and iii are (NO_3^-)-LDH, (M-citrate)-LDH, and M-LDH, respectively

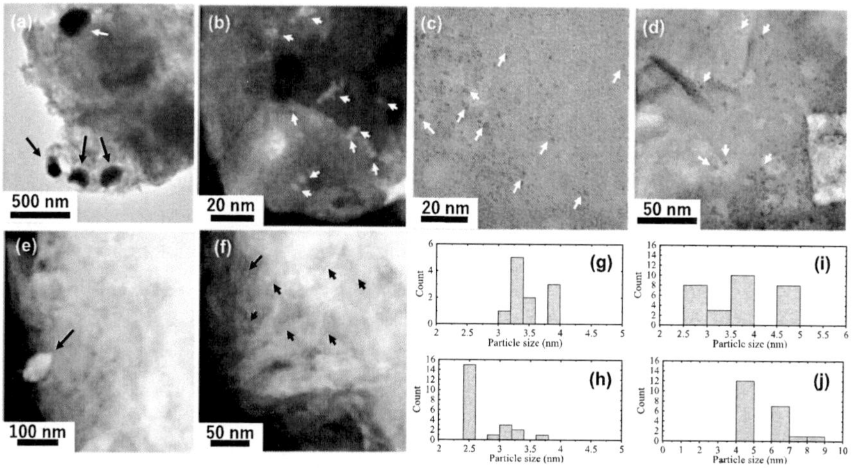

Fig. 3 TEM/STEM images of **a**, **b** Fe-LDH, **c** Co-LDH, **d** Ni-LDH, and **e**, **f** Cu-LDH with size distribution of **g** Fe, **h** Co, **i** Ni, and **j** Cu nanoparticles. Metal nanoparticles are pointed by white or black arrows

Table 1 BET surface areas and pore volumes of M-LDHs and (NO_3^-)-LDH

Catalyst	BET surface area m^2/g	Pore volume cm^3/g
(NO_3)-LDH	7.2	0.016
Fe-LDH	60.9	0.129
Co-LDH	67.7	0.116
Ni-LDH	66.1	0.129
Cu-LDH	59.2	0.153

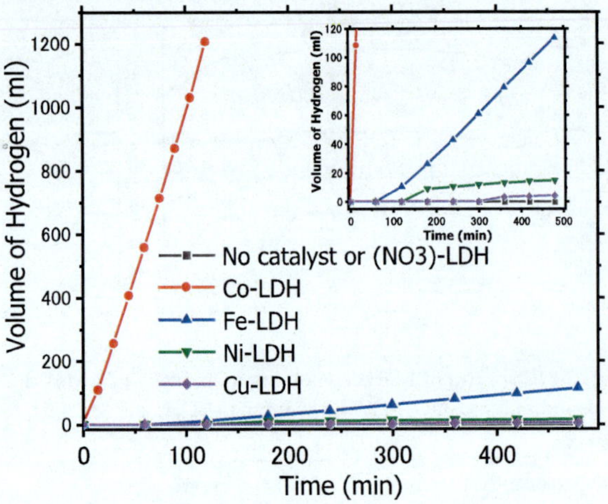

Fig. 4 Hydrogen production properties of (NO_3^-)-LDH and M-LDHs. Result of non-catalytic condition is also shown with black square

mol^{-1}) < Cu-LDH (72.2 kJ mol^{-1}). The value of activation energy of Co-LDH is comparable to other Co-based catalysts, e.g. 57.8 kJ mol^{-1} of Co–B/C [7] and 55.76 kJ mol^{-1} of Co-B/bentonite [8]. Therefore, it was demonstrated that cobalt had the highest catalytic activity for $NaBH_4$ hydrolysis due to the lowest activation energy in the series of transition metals, which were iron, cobalt, nickel, and copper. This is consistent with the comparison of HGR of various metal salts reported by Demirci et al. [1]

4 Conclusion

In this report, a versatile method to synthesize various non-precious transition metal nanoparticle supported on MgAl-LDH is presented. This method involves intercalation of the metal citrate complex anion followed by chemical reduction with $NaBH_4$. Based on the results, below are the summary:

- This synthesis route was successful in synthesizing small metal cluster of sizes less than 5 nm for Co and Ni. For Cu and Fe supported on LDH however, particles larger than 100 nm were also precipitated out of the LDH aggregate.

- Co nanoparticles supported on LDH had the highest hydrogen generation rate of 3725 ml min^{-1} $g_{Co}{}^{-1}$ followed by nanoparticles of Fe, Ni, and Cu supported on LDHs.
- Catalytic activity increased with temperature for all M-LDH. The activation energy, E_a of Co-LDH calculated from Arrhenius equation is 51.6 kJ mol^{-1}. This value is comparable to many other cobalt-based catalysts previously reported. This good catalytic activity may be attributed by the small and well distributed cobalt nanoparticle dispersed on the LDH structure.
- This work also provides a new idea for the development of a composite system consisting of LDH and non-precious metal nanoparticles for various other catalytic applications.

Acknowledgements This research was supported by the research grants provided from JSPS KAKENHI Grant Number JP21K05237.

References

1. U.B. Demirci, O. Akdim, J. Andrieux, J. Hannauer, R. Chamoun, P. Miele, Sodium borohydride hydrolysis as hydrogen generator: issues, state of the art and applicability upstream from a fuel cell. Fuel Cells **10**, 335–350 (2010)
2. Y. Kojima, K. Suzuki, K. Fukumoto, M. Sasaki, T. Yamamoto, Y. Kawai, H. Hayashi, Hydrogen generation using sodium borohydride solution and metal catalyst coated on metal oxide. Int. J. Hydrogen Energy **27**, 1029–1034 (2002)
3. E. Keceli, S. Ozkar, Ruthenium(III) acetylacetonate: a homogeneous catalyst in the hydrolysis of sodium borohydride. J. Mol. Catal. A-Chem. **286**, 87–91 (2008)
4. H. Inokawa, K. Okamoto, A.B. Mahpudz, Y. Ohgi, R. Tomoshige, K. Kusakabe, Formation of cobalt clusters in layered double hydroxide. J. Ceram. Soc. Jpn. **129**, 175–180 (2021)
5. A. Mahpudz, S.L. Lim, H. Inokawa, K. Kusakabe, R. Tomoshige, Layered double hydroxide supported cobalt nanocluster: size control and the effect in catalytic hydrogen generation. E3S Web Conf. **287**, 02009 (2021)
6. A. Mahpudz, S.L. Lim, H. Inokawa, K. Kusakabe, R. Tomoshige, Cobalt nanoparticle supported on layered double hydroxide: effect of nanoparticle size on catalytic hydrogen production by $NaBH_4$ hydrolysis. Environ. Pollut. **290**, 117990 (2021)
7. J. Zhao, H. Ma, J. Chen, Improved hydrogen generation from alkaline NaBH4 solution using carbon-supported Co–B as catalysts. Int. J. Hydrogen Energy **32**, 4711–4716 (2007)
8. İ. Kıpçak, E. Kalpazan, Preparation of CoB catalysts supported on raw and Na-exchanged bentonite clays and their application in hydrogen generation from the hydrolysis of $NaBH_4$. Int. J. Hydrogen Energy **45**, 26434–26444 (2020)

Chloralkali and Hydrogen Generation from Produced Water

Ana Hasrinatullina Bt M Basri, Terath Kumar s/o Omporkas, Anusha Nagaih, M Faris B M Shah, and Tan Loo Sen

Abstract Produced Water comprises the largest by-product stream associated with oil and gas recovery in addition to its inconsistent volume variation. Apart from that, its characteristics also varies from one field to another. As the oil and gas production approaching its maturation years, Produced Water volume also increases exponentially. The conventional method in managing Produced Water is to treat the Produced Water to an acceptable host authority requirement. For example, in Malaysia, this is governed by Exclusive Economic Zone (EEZ) requirement or Department of Environment (DOE) depending on the location of discharge and its distance from onshore. There are two common Produced Water management in Malaysia, firstly is to treat and dispose into depleted wells or treatment using physical separation equipment coupled with a chemical treatment program prior to discharging safely to the water bodies. This generally incurs huge Operations and Maintenance (O&M) expenditure as well as extensive Environmental Monitoring and Reporting (EMR) exercise to safeguard the environment. Current approach focused on the quantitative management of Produced Water as waste or unwanted by-products from oil and gas production in upstream. Little attention was given to the characteristics of the Produced Water especially its valuable salt content. It was found that, there is a significant amount of Sodium (Natrium) and Chloride in Produced Water in Malaysia which most probably caused by water breakthrough from sea water which is used to Improve

Ana Hasrinatullina Bt M Basri (✉) · Terath Kumar s/o Omporkas · A. Nagaih · M Faris B M Shah
PETRONAS Carigali Sendirian Berhad, Kuala Lumpur, Malaysia
e-mail: hasrinatulina@petronas.com

Terath Kumar s/o Omporkas
e-mail: terathkumar@petronas.com

A. Nagaih
e-mail: anusha.nagai@petronas.com.my

M Faris B M Shah
e-mail: faris.shah@petronas.com.my

T. L. Sen
PETRONAS Gas Berhad, Kuala Lumpur, Malaysia
e-mail: loosen.tan@petronas.com.my

M. B. Othman et al. (eds.), *Proceedings of the 1st International Conference of New Energy*, Springer Proceedings in Energy, https://doi.org/10.1007/978-981-99-0859-2_17

Oil Recovery (IOR). This has sparked the interest in exploring the opportunity to utilize Produced Water to generate Chloralkali chemicals. Produced Water which used to be viewed as waste would now to be perceived as a valuable commodity. The advantage of Chloralkali generation from Produced Water against the conventional method is that there is no requirement of adding additional salt to the sea water. The ultimate objective of this study would be to maximize the utilization of Produced Water and potential introductory of a new stream of revenue for PETRONAS. There are two well-known methods to produce Choralkali namely electrolysis or membrane cell technology. The variation of technology also produces a different kind of Choralkali products such as NaOH, Cl_2, or H_2 or NaOCl and H_2. For this experiment, the electro-chlorination unit producing NaOCl and H_2, a readily off-the-shelf technology has been explored.

1 Background/ Problem Statement

All industries have to manage their unwanted by-products whether in a form of emission, liquid discharge, or solid waste generation. Usually, these by-products managed as per regulatory requirement without noticing their locked-in-potential. In mature oil and gas production the largest by-products is Produced Water. Produced Water comes with variety of quality which depending on reservoir characteristics; some can be environmentally safe while at some locations, the Produced Water can contain hazardous contaminants such as Mercury, Arsenic, Lead or at times it is acidic. In Malaysia, this is governed by Exclusive Economic Zone (EEZ) requirement or the Department of Environment (DOE) depending on the location of discharge and its distance from onshore. There is also PETRONAS Produced Water Commitment enforced through Malaysia Petroleum Management (MPM) Minimum Environmental Standard (MES). For such a case where Produced Water is unsafe then the treatment before discharge is compulsory to meet to safe in addition to its regulatory enforcement. This generally incurs huge Operations and Maintenance (O&M) expenditure as well as extensive Environmental Monitoring and Reporting (EMR) exercise to safeguard the environment (Fig. 1).

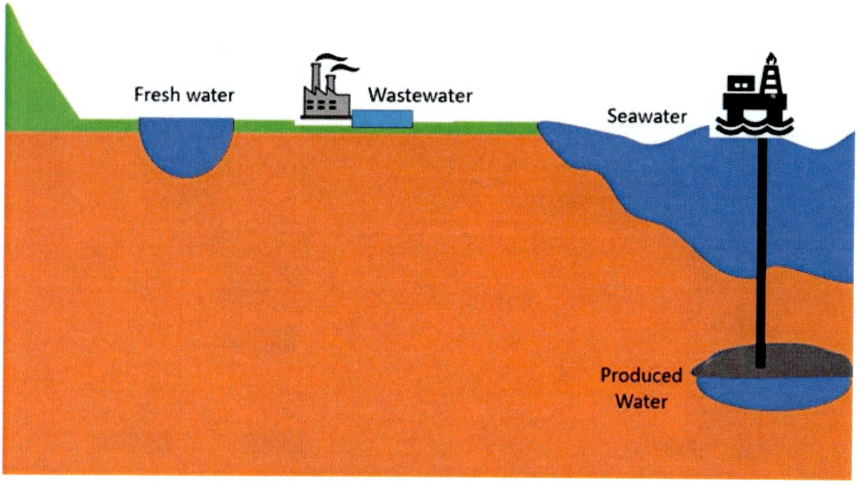

Fig. 1 Source of produced water

2 Method

The key to this study is to evaluate by-products circularity based on World Business Council for Sustainable Development (WBCSD) [1] Circular Transition Indicator (CTI). This starts with scanning samples of Produced Water from various locations for precious minerals content. Some of the samples revealed that there are significant amount of Sodium and Chloride exist in Produced Water in Malaysia. This later triggers assessment of potential Chloralkali generation. Hence, a laboratory trial was initiated.

Later in a laboratory experimental project, samples of Produced Water were tested using a lab-scale/portable electro-chlorination unit which principally operates on electrolysis. Produced Water electrolysis generates Sodium Hypochlorite (NaOCl) also known as Chloralkali chemical as primary product and Hydrogen (H_2) as by-product. Handling such delicate process certainly requires extra safety precaution as the by-products are hazardous in nature. The encouraging results from this lab test ensured project doability of Cloralkali generation from Produced Water. Later, the samples were sent for analysis to quantify the amount of free Chlorine and total Chlorine present in the water which will determine the effectiveness of electro-chlorination process (Fig. 2).

The electrolysis process takes place as below:

- Anode: $2\ Cl^- \rightarrow Cl_2 + 2e^-$
- Cathode: $2\ Na^+ + 2\ H_2O + 2e^- \rightarrow 2\ NaOH + 2\ H_2$
- Overall: $2\ NaCl + 2\ H_2O \rightarrow 2\ NaOCl + 2\ H_2$

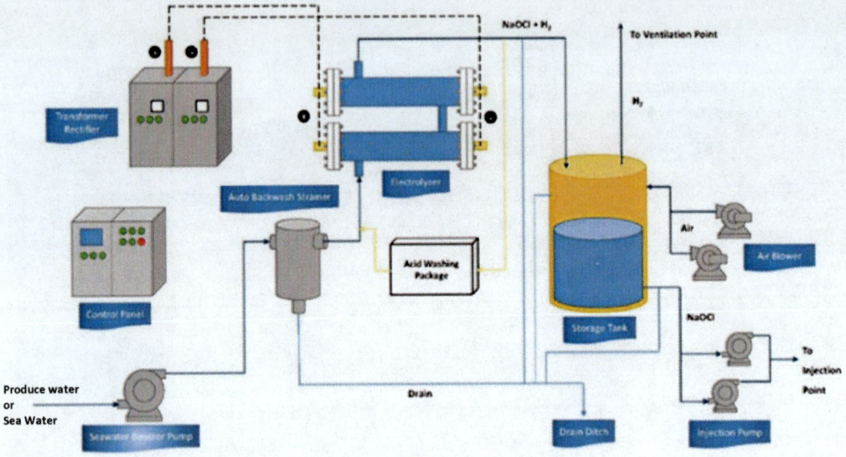

Fig. 2 A typical full-scale land-based electro-chlorination unit [2]

3 Results

The positive outcome from the lab experiment demonstrated and validated that the chosen method of using electro-chlorination unit to generate Chloralkali from Produced Water is effective. From this experiment, the amount of free Chlorine present was more than 20 ppm while the total Chlorine present is more than 500 ppm. Presence of Chlorine indicated that the conversion took place hence Chloralkali generation is highly possible.

Moreover, throughout an economic analysis in a real commercial implementation, generation of Chloralkali from Produced Water using electro-chlorination unit could reduce the operating cost and contribute to an annual OPEX saving estimated at RM200,000 per year. The calculation basis is made by referring to one of Operating Petronas Unit (OPU) as a bench head segment. In addition, Hydrogen gas which is a by-product from the generation of Chloralkali via electro-chlorination unit can be further explored as a raw material to other commercial studies such as to be reacted with inherent Carbon Dioxide from acid gas removal unit in order to generate green methane or low Carbon Methanol. The technical and economic study was conducted by assessing the supply and demand of Chloralkali for Petronas own-consumption or potential marketability of the products (Fig. 3).

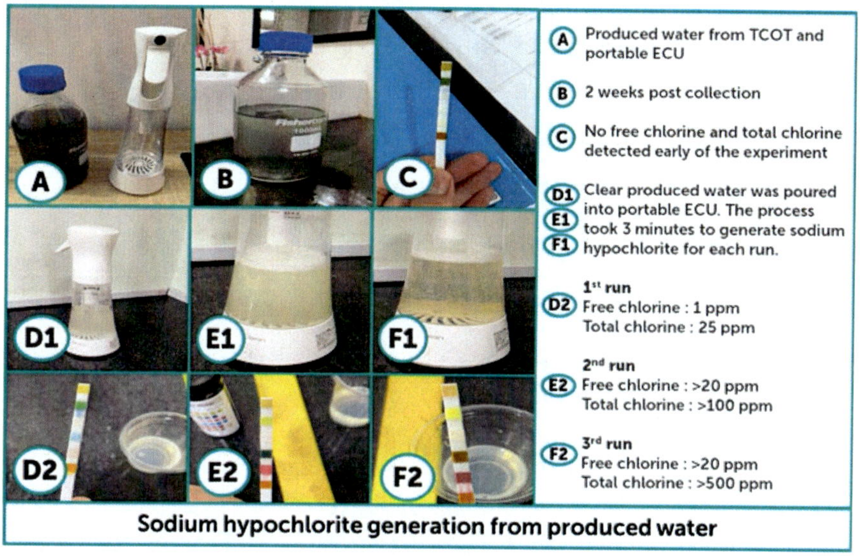

Fig. 3 Laboratory experiment result from Terminal "X"

4 Novel Information

Current challenges of increasing Produced Water are common across oil and gas producers. The intricate characteristic of Produced Water is influenced by different concentrations of organic and inorganic compounds which requires customized innovative solution. Full characteristic of Produced Water must be conducted to identify salt recovery potentials even for direct usage such as irrigation water. In depth understanding of supply chain, this enables cross-industries collaboration and sharing of resources to meet industrial supply and demand.

The conventional Chloralkali production practice Sodium Chloride dilution in fresh water or generated from treated sea water. There are processes that use molten Sodium Chloride to produce Chlorine and Sodium metal or condensed hydrogen Chloride to produce Hydrogen and Chlorine. This process has a high energy consumption, around 2,500 kWh (9,000 MJ) of electricity per ton of Sodium hydroxide produced. The Chloralkali chemicals selling price also influenced by the price of Sodium Chloride salt, fresh water, and sea water extraction cost. The purity of Sodium Chloride salt is also a concern for quality control. There are limited studies on salt separation from Produced Water nor any commercial application that could be used as reference. Chloralkali is listed as top 10 [3] industrial chemicals mostly produced worldwide and always on demand especially for water treatment (Fig. 4).

There is a huge demand for Chloralkali which principally used as intermediate raw materials—as sanitizing agents, in pulp and paper processing, soap, detergent, and many more. This initiative is highly feasible and supports circular economy and is ready for mass production deployment. Nevertheless, the related cost of product

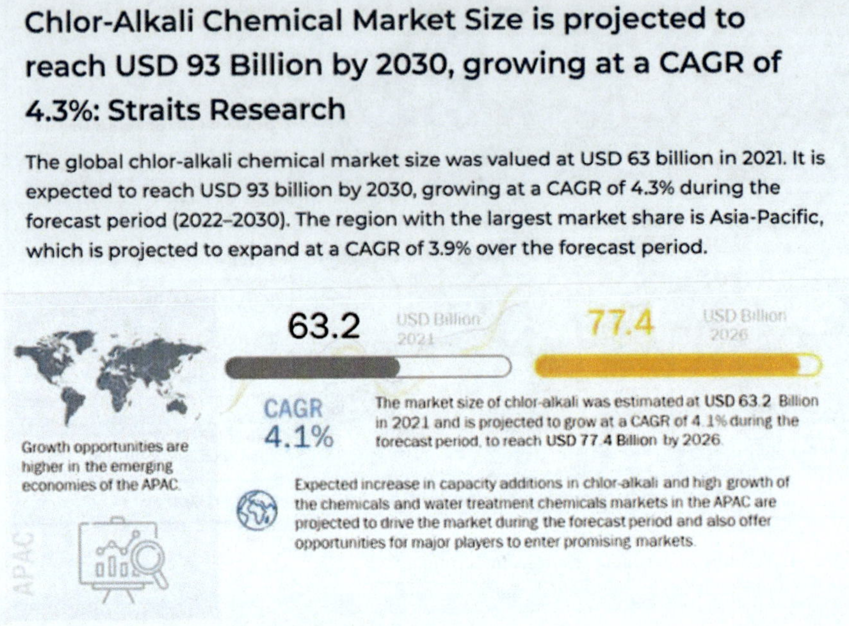

Fig. 4 Attractive opportunity in Chloralkali market [3]

transportation and delivery shall be considered diligently as it is costly to store and deliver Chloralkali chemicals safely to customers.

5 Challenges and Opportunities

Produced Water has lower and inconsistent amount of salt compared to the conventional method of sourcing salt and fresh water to get a consistent brine solution for generating Chloralkali. This will also affect the amount of hydrogen produced as by-products. Presence of other contaminants affecting general stoichiometric balance of the solution jeopardizing the final product quality. Chloralkali is in liquid form and easily transported. However, care must be made for Hydrogen storage and transportation which possibly can be managed through carbon circularity. Production cost for hydrogen through this process is almost negligible.

6 Conclusion

The Circularity Gap Report 2022 [4] indicates that we only cycle 8.6% of what we use, which leaves a massive resource circularity gap of over 90%. It is convenient to continue oil and gas industry the way it was 200 years ago, but if there is a sustainable way for the future; this path must be urgently explored. These days, more locations are identified with freshwater scarcity areas hence requiring industries to shift the demand of freshwater for industrial use to more sustainable source. Furthermore, technology for Produced Water treatment and effluent treatment to meet most stringent requirement has become more affordable with a better awareness on sustainability could support the journey toward circularity (Fig. 5).

The 2030 Agenda for Sustainable Development adopted by all United Nations Member States in 2015 provides a shared blueprint for peace and prosperity for people and the planet, now and into the future. At its heart are the 17 Sustainable Development Goals (SDGs), whereby the generation of Chloralkali from Produced Water specifically supports Sustainable Development Goal 12 (SDG12) "Responsible Consumption and Production" to reduce waste generation and recycling the by-product to add value to circular economy growth especially across industries (Figs. 6 and 7).

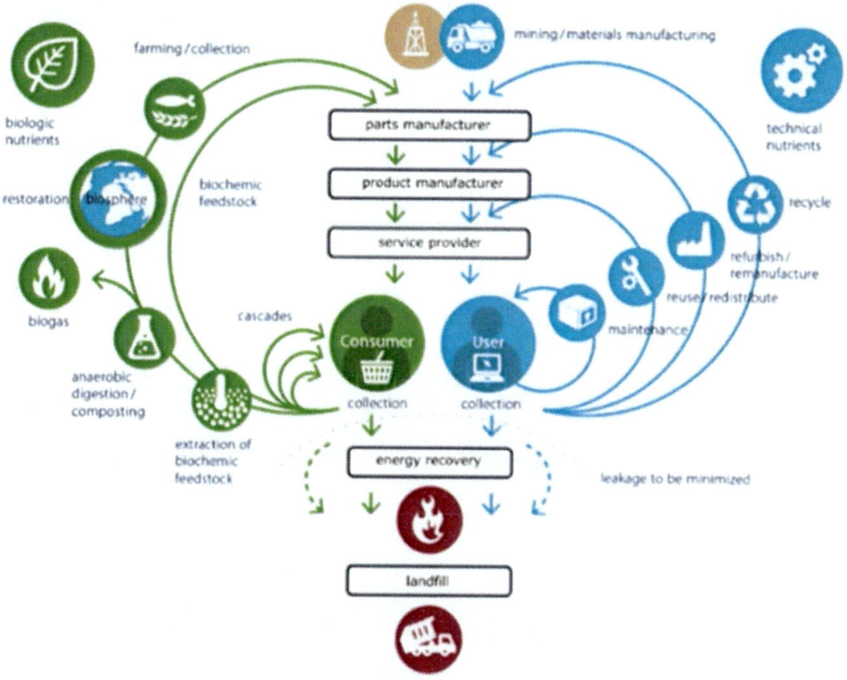

Fig. 5 Ellen MacArthur Foundation [5], circular economy system

Fig. 6 United Nation [6], 17 sustainable development goals adopted in 2015

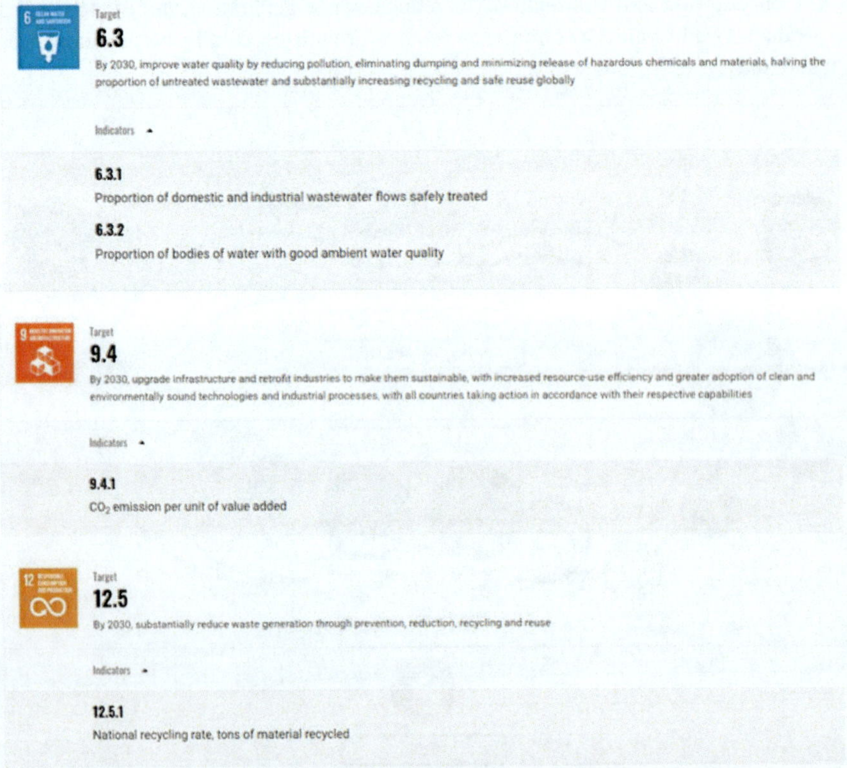

Fig. 7 Sustainable development goals extract from 6, 9 and 12 from the list 17

Acknowledgements The authors would like to thank Mazri B Mohd Ali—Head, Operational Excellence, PETRONAS Operation team; Joint Collaboration Team of Generation of Chloralkali and Hydrogen from Produced Water; Group Research & Technology (GR&T) for conducting this study which is in line with PETRONAS Statement of Purpose "*Progressive energy and solution partner enriching lives for sustainable future*".

References

1. World Business Council For Sustainable Development (WBCSD). Accessed 2022/10/31
2. Industrial Today, http://www.industrialtoday.co.uk. Accessed 2022/10/31
3. GlobeNewswire, https://www.globenewswire.com. Accessed 2022/11/31
4. Circularity Gap Report 2022, http://circulareconomy.europa.eu. Accessed 2022/10/31
5. Ellen Macarthur Foundation, http://www.ellenmacarthurfoundation.org. Accessed 2022/10/31
6. United Nation, http://sdgs.un.org/goals. Accessed 2022/10/31

Ammonia as a Hydrogen Vector: Validated Large Eddy Simulation of Ammonia Co-Firing in a Pilot-Scale Coal Combustor

**Mohammad Nurizat Rahman,
Muhamad Shazarizul Haziq Mohd Samsuri, Suzana Yusup,
and Ismail Shariff**

Abstract It is anticipated that ammonia (NH_3) will be used as a carbon-free substitute for coal. Even so, regulating NH_3-coal co-firing while preserving both nitrogen oxides (NO_x) emission and flame stability is an important concern that necessitates the use of a dependable technique to aid in the combustion tuning of the said co-firing. Computational fluid dynamics (CFD) has the capacity to be a promising method for assisting in the said co-firing tuning for NO_x reduction in the power plant. Hence, the validated numerical model of NH_3 co-firing with sub-bituminous coal was established in the current study through a detailed Large Eddy Simulation (LES) using a pilot-scale coal combustor as the main geometry basis. The predictive performance of the CFD model was assessed by comparing it to actual experimental data from the said coal combustor facility, which revealed a less than 10% difference in predicted NO_x and temperature profiles for both pure coal firing and NH_3 co-firing. Therefore, the validation can be considered satisfactory, and the model can predict NO_x emission and combustion behaviour with acceptable accuracy for both coal and NH_3 co-firing. Overall, the validated CFD model can be used to gain in-depth insights into NH_3 co-firing to aid in combustion tuning for future co-firing implementation at actual utility plants.

Keywords Ammonia co-firing · Coal: CFD · LES · Emission · Combustion · Power plant

1 Introduction

Presently, the electricity sector makes up roughly 40% of global carbon dioxide (CO_2) emissions, and electricity demand is predicted to rise by more than 50% by 2040 [1, 2]. It is well-known that coal-fired thermal power plants account for a sizeable portion of the world's primary electricity production [2–5] because of its abundant reserves and competitive pricing [6]. Since coal fuel has a high carbon content [8]

M. N. Rahman (✉) · M. S. H. M. Samsuri · S. Yusup · I. Shariff
Generation Unit, Generation and Environment, TNB Research, 43000 Kajang, Selangor, Malaysia
e-mail: izatfariz49@gmail.com

M. B. Othman et al. (eds.), *Proceedings of the 1st International Conference of New Energy*, Springer Proceedings in Energy, https://doi.org/10.1007/978-981-99-0859-2_18

and is one of the main anthropogenic sources of CO_2 emissions [2, 4, 9], coal-fired thermal power plants produce the most CO_2 of any type of power generation facility [7], accounting for 30.4% of global CO_2 emissions in 2018 [10]. As mitigating CO_2 emissions is now becoming a global agreement, reduced greenhouse gas emissions are essential for managing climate change brought on by global warming [7, 9–13]. The goal of a low-carbon society is widely discussed, and coal-fired utilities are under increasing pressure to decarbonise [5].

In order to reduce CO_2 emissions from the aforementioned power plants, a number of techniques are being developed, including ultra-supercritical technology [10], integrated gasification combined cycle (IGCC) [10], double reheat technology [10], carbon capture and storage (CCS) [10], oxy-fuel combustion [7], carbon capture and storage [7], and the use of low-carbon/carbon–neutral fuels [10, 11].

Biomass and biogas are appealing carbon–neutral fuels for co-firing when it comes to using low-carbon/carbon–neutral fuels [5, 14], but seasonal variations in feedstock supply present significant challenges [5]. Hydrogen, a different form of carbon-free fuel, is also anticipated to be crucial in the development of a low-carbon society in the future [9, 11–13, 15]. However, due to its high volatility [11], transporting and storing hydrogen remains challenging [13, 16]. One of the efficient substitutes for hydrogen carriers shown in a study by Rahman [13] is ammonia (NH_3), which has a very high hydrogen density compared to liquid hydrogen and organic hydride [3, 7, 12, 13, 16, 17]. The fact that NH_3 retains about 90% of the energy present in the hydrogen feedstock and is much easier to liquefy and store makes it a desirable hydrogen vector [18]. Having said that, NH_3 should be seen as an enabler rather than a rival to the hydrogen economy [19].

A recently proposed method for reducing CO_2 emissions from coal-fired power plants is co-firing coal with NH_3 [4, 7, 10, 12, 13, 17]. This is because co-firing maximises the use of the facilities already in place in the numerous coal-fired power plants, reducing resource waste and financial loss from the early retirement of power plants [20]. Additionally, merely lowering the carbon content of the fuel stream could hasten the reduction of CO_2 emissions [9].

However, due to its significantly higher fuel-nitrogen content, co-firing NH_3 in coal-fired boilers may result in increased nitrogen oxides (NO_x) emissions [9, 12]. The combustion properties of NH_3 have been the subject of research by numerous organisations worldwide [7]. The study of combustion dynamics and the creation of thorough reaction mechanisms are aided by useful data on the propagation of NH_3 flames under various conditions [12, 16]. There have been recent developments in the study of NH_3 co-firing technology with pulverised coal [3, 20]. IHI Corporation tested NH_3 co-firing with a 20% NH_3 co-firing ratio at a pulverised coal combustion facility (10 MW) with success [3]. Additionally, the experimental findings showed that, with the right NH_3 injection technique, NO_x emission and unburned carbon (UC) in fly ash could be comparable to coal firing in the case of NH_3 co-firing [3]. In a different experimental study, a horizontal single burner with a coal feeding rate of 100 kg/hr was used to explore the co-firing characteristics of pulverised coal and NH_3 [3]. Their findings revealed that when NH_3 was injected from the centre of the burner with NH_3 of 20 cal.%, NO_x concentration in the flue gas increased by about

20% and UC in fly ash increased slightly [3]. According to these findings, controlling NH_3-coal co-firing while maintaining NO_x emission and flame stability is a critical challenge that calls for additional study in order to achieve the desired combustion characteristics.

Numerous significant studies on NH_3 co-firing only used an ideal reactor network model [4, 5, 7, 9, 12], which is known to ignore fluid dynamics and 3D effects. Although the energy density of NH_3 and low-rank coals is comparable [19], it is obvious that important NH_3 characteristics, such as low energy and laminar burning velocities [16, 21], high auto-ignition temperatures, and elevated ignition energies [19], will make it more challenging to co-combust NH_3 optimally and achieve the combustion performance required by coal-fired power plants. In fact, more research is required before NH_3 can be used in actual coal-fired power plants in order to better understand its emission and combustion properties. In particular, more research that takes into account the fluid dynamics and kinetics aspects of the NH_3 co-firing to help with flow tuning is required.

Detailed analyses of NH_3 co-firing could well be provided using computational fluid dynamics (CFD), which has the potential to be a practical technique. In utility furnaces that burn pulverised coal, CFD has frequently been used to examine heat transmission and combustion characteristics [3]. As a result, in this study, a detailed Large Eddy Simulation (LES) was used to establish a validated numerical of NH_3 co-firing based on a pilot-scale coal combustor facility. Because sub-bituminous coal rank is the most common coal rank used in Malaysia's power plants [22, 23], it was used in the numerical assessments. The model's ability to predict NO_x concentrations and combustion profiles was evaluated by comparing it with actual experimental data from the aforementioned coal combustor facility.

2 Experimental Setup

The coal combustor facility at TNB Research is where the NH_3 co-firing took place, as shown in Fig. 1. The co-firing assessment uses a thermal input of 150 kW. In order to properly mix with the coal-air mixture, the NH_3 jet velocity was lowered and the NH_3 was injected concentrically and radially in the centre of the coal injector burner. This allowed direct mixing with the incoming coal and air. A single swirl burner, K-type thermocouples at the combustor sections, and a flue gas analyser are all features of the coal combustor facility. The heat exchanger, cyclone, main combustion area, and air and fuel inlet are the combustor's main parts.

The combustor is set up in an L-shape to mimic a boiler layout, making it possible to distinguish between the convection and radiation (high temperature combustor/flame zone) areas. While the secondary air (SA) acts as an oxidising agent during the combustion process, the primary air (PA) acts as a transport medium for the coal sample. The PA volume flowrate was maintained at 9 Nm^3/hr throughout the test programme, whereas the coal, SA, and NH_3 volume flowrates were determined based on calorific/thermal input. A total of four tube sections measuring 3.3 m in

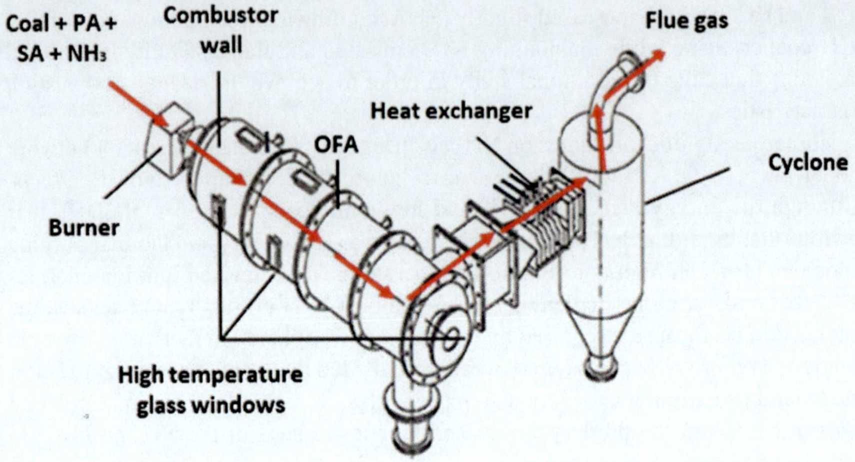

Fig. 1 Coal combustor facility at TNB research

length and 0.6 m in internal diameter make up the combustor's main body (longer length). One K-type thermocouple is included in each tube section to measure the temperature as it operates. Sections 2, 3, and 4 come after Sect. 1 (upstream region after the inlets). The internal diameter of the downstream portion of the combustor, which is the shorter length, is 0.3 m, and its length is roughly 0.91 m.

At the combustor outlet, both the gas temperature and the NO_x concentration were continuously monitored. Liquefied petroleum gas (LPG) was continuously fed with a fixed mass flowrate of 2 kg/hr throughout the test programme to maintain a high-temperature condition within the combustion zone and simulate the actual boiler temperature condition. Since it is prohibited to release NH_3 into the surrounding air due to its toxicity, the NH_3 in the flue gas was also measured to ensure that no NH_3 slip occurred [20]. One case of coal firing and one case of NH_3 co-firing with a 60 cal.% proportion were both conducted. In order to replicate the air staging system used in real power plants to reduce NO_x, an air staging ratio, which is the fraction of over-fire air (OFA) of about 20% was applied between Sects. 2 and 3 of the combustor. Table 1 displays the fuel properties used in the experiments and numerical studies.

Table 1 Properties of fuels

Proximate analysis, wt.%, ad., coal (TM-total moisture, VM-volatile matter, FC-fixed carbon, AC-ash content)				Ultimate analysis, wt.%, ad., coal (C-carbon, H-hydrogen, N-nitrogen, O-oxygen, S-sulphur)					GCV-gross calorific value, ad., coal (kcal/kg)		
TM	VM	FC	AC	C	H	N	O	S	Coal	NH_3	LPG
23.80	40.98	39.00	2.40	67.70	4.41	0.89	23.67	0.17	6449	5374	11,775

3 Numerical Setup

The combustion processes of coal firing and NH_3 co-firing in the coal combustor facility were simulated using CFD techniques. The software ANSYS FLUENT 19.0 was employed. The numerical model utilised the reacting and compressible Navier–Stokes (NS) equations. To resolve the governing equations, the pressure-based solver was employed. Using the LES approach, turbulence was solved [24]. On the formulations used in the NS equations and LES model for coal combustion, Wan [25] and Sun [26] provide in-depth information. Coal devolatilisation, char conversion/reaction, and volatiles reactions are the three main stages of the coal combustion process that were taken into account in the coal firing simulation using the same sub-models and kinetics reactions as NH_3 co-firing. In order to determine the volatile composition and rate constants for coal devolatilisation, the advanced coal network model and the coal database from our own analytical fuel laboratory were used.

Our earlier research [22, 23] demonstrates the detailed chemical kinetics and coal combustion models used in the most recent CFD evaluations. Second order upwind methods were used to discretise each equation. A Lagrangian method that considered turbulent dispersion was used to trace the trajectories of coal particle. The discrete ordinate (DO) method, with an angular discretisation of 5 divisions and 3 pixels in both the polar and azimuthal orientations, was used to predict radiation. Utilising the weighted-sum-of-gray-gases model (WSGGM) [3], the gas emissivity was calculated.

The production of NO_x was simulated using a post-processing technique. To start, temperature, major gas composition, and velocity distributions were derived using combustion simulations. The combustion computation was then used to incorporate the reactions of NH_3, hydrogen cyanide (HCN), thermal NO_x, and NO_x reduction by char. The equations for flow, turbulence, other major gas compositions like oxygen, CO_2, carbon monoxide (CO), and hydrogen, energy, and radiation were not solved. Instead, only NO_x-related species such as NO, NH_3, HCN, O, hydroxide (OH), and N were computed.

4 Grid-Convergence Analysis

Using the Assembly Meshing technique, the computational domain for coal combustor was discretised with hexahedral-dominant meshes. To ensure that the prediction outcomes are unaffected by the quantity of meshes produced, a mesh independent test is carried out. The characteristics of the various meshes used in the computational domain are highlighted in Table 3. Meshes are made with orthogonal quality and skewness taken into account to reflect mesh quality because mesh quality affects the degree of spatial discretisation error [27]. How closely the angles between successive mesh faces resemble the ideal angle is indicated by the orthogonal quality.

Table 3 Characteristics of the various meshes being used

Average element size (mm)	Grid points (10^6)	Meshes (10^6)	Orthogonal quality	Skewness
80	0.036	0.030	0.916	0.094
40	0.043	0.038	0.936	0.079
20	0.123	0.116	0.971	0.051
10	0.786	0.767	0.990	0.024
8	1.734	1.687	0.990	0.015
5	6.091	6.024	0.996	0.019

The highest quality on the orthogonal quality scale, which ranges from 0 to 1, is 1 [27].

On the other hand, the skewness shows how close the mesh is to the ideal equiangular mesh. Because the governing equations are solved under the assumption that the meshes are largely equiangular, heavily skewed faces and meshes are unsuitable. A value close to zero has the smallest deviation from a normalised equiangular angle, and skewness ranges from 0 to 1 [27]. All generated meshes were optimised to produce high-quality average orthogonal and average skewness, as shown in Table 3.

Figure 2 depicts the anticipated NO_x (coal firing case) as the mesh count varies in the coal combustor. With a variation of less than 1%, the NO_x hardly ever changes as the mesh number is raised from 1.687 million to 6.024 million. 1.687 million meshes are therefore chosen for the coal combustor model.

The independent mesh model of the coal combustor computational domain is shown in Fig. 3. The upstream region (inlet) has a finer mesh than the downstream domain area since it involves complex mixing and reactions of coal, NH_3, LPG, and air.

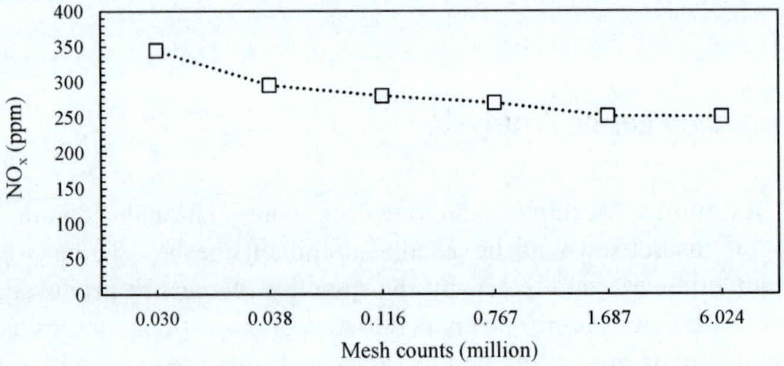

Fig. 2 Predicted NO_x at varying mesh counts

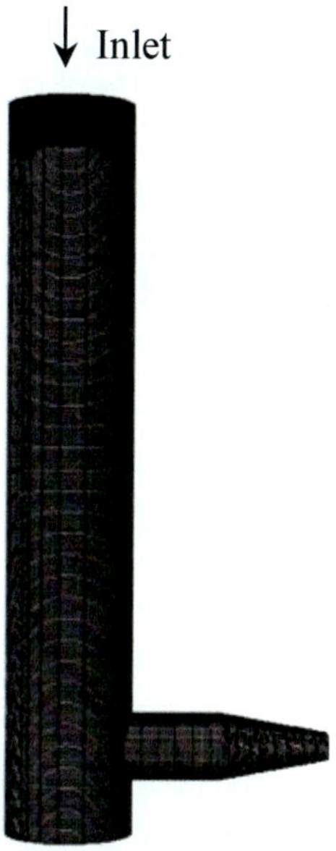

Fig. 3 Coal combustor mesh model

5 Results and Discussion

Validation I: NO_x results

In both experiment and numerical simulation, NH_3 co-firing produces higher NO_x concentrations at the combustor exit compared to pure coal firing, as shown in Fig. 4. The results of the experiments show that the NO_x concentrations from coal firing and NH_3 co-firing are 265 and 1573 ppm, respectively. Therefore, the NO_x concentration increases by about 494% as the NH_3 fraction rises to 60 cal.%, which is roughly 5 times more than the NO_x concentration produced by coal combustion. As shown in the experimental results in Fig. 4, while reducing thermal and prompt NO_x is always challenging in carbon-based combustion, adding NH_3 further increases the likelihood of fuel-bound nitrogen interacting with oxygen. Total NO_x, which includes thermal, prompt, and fuel NO_x, significantly increases as a result of the synergistic effect of the fuel-bound nitrogen.

The model appears to perform well when compared to experiment results, with NO_x concentration discrepancies between the experiment and the prediction for both coal firing and NH_3 co-firing being less than 10%. Although the current model was able to roughly predict the same NO_x increment trend when NH_3 co-firing was used, there is still a small difference between the numerical model's NO_x predictions and those found in experiments. This may be explained by the fact that the NO_x modelling employed is based on a semi-empirical mechanism [3]. The NO_x production reaction pathways are actually much more complex than they appear to be, which makes it

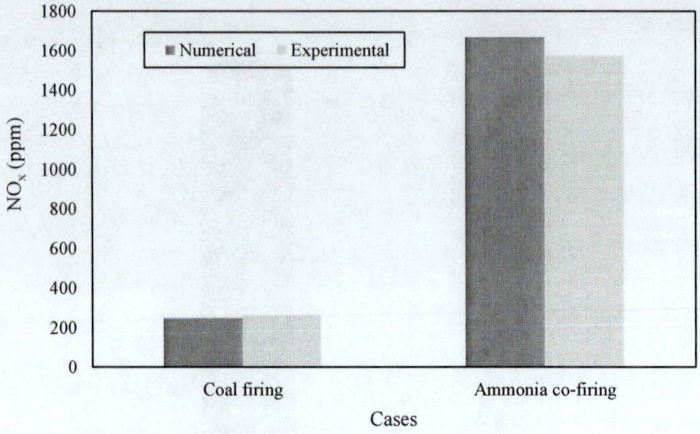

Fig. 4 NO_x data from experimental and numerical assessments

more expensive to compute the actual kinetics involved [12]. As a result, semi-empirical mechanisms, which are great for scaling and parametric analysis, might not give precise NO_x values. However, the simulation cases depicted in Fig. 4 showed a NO_x difference of less than 10% when compared to the experimental data. Hence, the quantitative trend seen in the experiment data is consistent with the trend in NO_x concentrations predicted by the numerical model. Due to the model's ability to predict NO_x emission with acceptable accuracy for both coal firing and NH_3 co-firing, the validation can be deemed adequate.

It is also important to note that during the experimental testing, no NH_3 slip was found at the combustor outlet. As a result, the testing has fully addressed the safety concern of the dangerous NH_3 slip. The NH_3 injection direction in the coal combustor is responsible for the lack of NH_3 slip. As mentioned before, NH_3 was injected concentrically and radially in the middle of the coal injector burner, allowing for quick mixing with the incoming coal and air and reducing the NH_3 jet velocity to allow for better mixing with the coal-air mixture. In addition to improving mixing, it reduces the likelihood that NH_3 will infiltrate the primary firing area of the coal combustor's recirculation zone. Results from a prior study, which showed that high NH_3 axial jet velocity can penetrate the recirculation zone [7], also support this. As a result, NH_3 directional injection plays a crucial role in reducing the likelihood of NH_3 slip.

Validation II: Temperature profiles

With regard to Fig. 5, thermocouples were used to measure the actual temperature in each tube section of the aforementioned coal combustor facility. According to the experimental data, the average temperature difference for the coal firing and NH_3 co-firing cases ranges from 0.9 to 6.1%. The average difference in the numerical data for both of the aforementioned cases ranges from 0.4 to 4.9%. The temperature differences have been found to be negligible in both coal firing and NH_3 co-firing

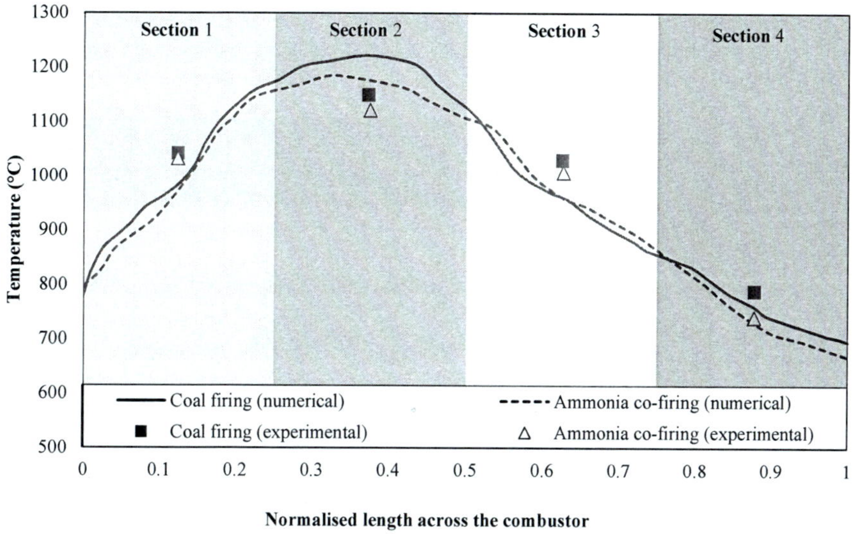

Fig. 5 Temperature data from experimental and numerical assessments

cases due to the constant thermal input applied to both cases. The section of the combustor with the highest temperature according to experimental data was Sect. 2, which is also the section with the highest temperature according to numerical assessments. Due to the enrichment of OFA between Sects. 2 and 3, which serves as a thermal NO_x mitigation strategy and halts the flame front before it reaches Sects. 3 and 4 for safety reasons, the peak temperature in the combustor has experienced a significant drop after Sect. 2.

The maximum temperatures determined from the experimental data are 1149 and 1121 °C for coal firing and NH_3 co-firing, respectively. This means that the peak temperature percentage difference between these two cases, as determined by experimental data, is 2.4%. The maximum temperatures that the numerical model predicts for coal firing and NH_3 co-firing are 1222 and 1184 °C, respectively. Therefore, according to numerical data, there is a 3% difference in the peak temperature between these two cases. Peak temperature differences between experimental and numerical data are 5.9 and 5.3% for coal firing and NH_3 co-firing cases, respectively.

The computation of the turbulence-chemistry model in the simulation is one of the potential causes of the small difference. Despite the fact that the turbulence-combustion model used in the current study offers one of the best possible balances between the accuracy and efficiency of the numerical simulation, the LES algorithm filters the relatively small eddies to be modelled by the sub-grid formulation [24]. Therefore, the current model does not adequately represent the diverse range of length scales. This might lead to minimal regional variations in post-combustion temperatures and flame front dynamics.

The difference in peak temperature between experimental and numerical data is, however, only less than 6%. Additionally, the numerical assessment's predictions for

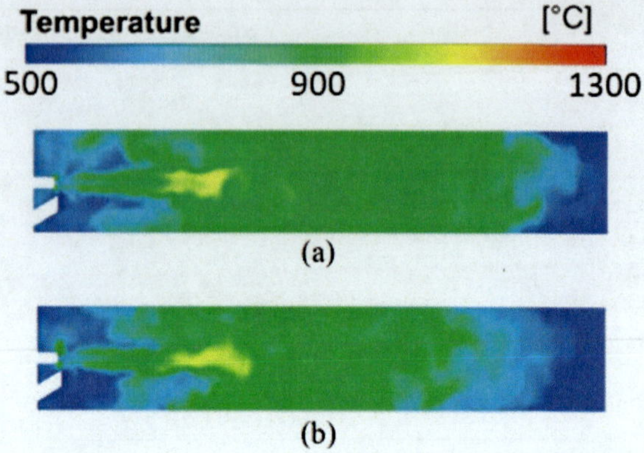

Fig. 6 Temperature contours for **a** coal firing and **b** NH_3 co-firing

the positioning and trend of the temperature profiles closely match those obtained from experimental data. As a result, the validation can be deemed adequate, and the model can predict temperature behaviour for both coal firing and NH_3 co-firing with decent accuracy. Figure 6 depicts the temperature contour for both cases as predicted by the numerical model, demonstrating that the validated model can also offer in-depth interpretations of the flame structure produced.

6 Conclusion

It is expected that using NH_3 as a carbon-free coal substitute, at least in part, will help with the gradual phase out of coal and reduce CO_2 emissions from pulverised coal-fired boilers. In this study, the LES formulation was used to establish the CFD model for NH_3 co-firing with coal. The main objective of developing a validated model is to use it as one of the tools to provide reliable insights on the combustion behaviour of NH_3 co-firing for a future industrial implementation of the said co-firing.

The model's predictive ability was first assessed by comparing it with actual experimental data from the aforementioned coal combustor facility, which showed a difference in predicted NO_x for both pure coal firing and NH_3 co-firing of less than 10%. Then, the model's predicted temperature profile for the coal combustor was compared to temperature readings from the actual coal combustor. With a difference in average temperature between the model's predictions and the experimental data of less than 10%, it has been discovered that the model appropriately predicts the actual temperature trends in the combustor. Furthermore, the predicted peak temperature location closely matched the ones found in the experimental work. In light of this, the validation can be considered adequate, and with reasonable accuracy, the model can predict temperature behaviour for both coal firing and NH_3 co-firing.

The current work, however, simplifies the combustion zone and does not thoroughly examine the effects of having multiple burners. Additionally, the coal combustor used in this study is considerably smaller than a typical coal-fired utility furnace. Due to complexity, such as the presence of flame interaction or the influence of residence time, it is expected that estimating the amount of emissions will become more difficult when multiple burners are used, as in commercial boilers. It is possible to conduct additional research to examine the effects of various NH_3 burner designs and multiple burners on emission and combustion characteristics.

Acknowledgements The authors would like to acknowledge their gratitude to TNB Research Seeding Fund grant number TNBR/SF 429/2022 for the funding of the project.

Authorship Contribution Statement **Mohammad Nurizat Rahman**: Principal investigator, Conceptualisation, Writing—original draft, Writing—review and editing, Visualisation, Lead CFD and combustion analyst.

Muhamad Shazarizul Haziq Mohd Samsuri: Lead experimental.

Suzana Yusup: Supervision.

Ismail Shariff: Supervision.

Statements and Declarations Research fund: TNB Research Seeding Fund grant number TNBR/SF 429/2022.

References

1. Z. Cesaro, M. Ives, R. Nayak-Luke, M. Mason, R. Banares-Alcantara, Ammonia to power: forecasting the levelized cost of electricity from green ammonia in large-scale power plants. Appl. Energy. (2021)https://doi.org/10.1016/j.apenergy.2020.116009
2. J.S. Cardoso, V. Silva, J.A.M. Chavando, D. Eusebio, M.J. Hall, Numerical modelling of the coal phase-out through ammonia and biomass co-firing in a pilot-scale fluidized bed reactor. JFUECO (2022). https://doi.org/10.1016/j.jfueco.2022.100055
3. J. Zhang, T. Ito, H. Ishii, S. Ishihara, T. Fujimori, Numerical investigation on ammonia co-firing in a pulverized coal combustion facility: effect of ammonia co-firing ratio. Fuel (2020). https://doi.org/10.1016/j.fuel.2020.117166
4. N. Tsukada, N. Kinoshita, Y. Kabuki, Y. Taguchi, Y. Takashima, T. Tsumura, M. Taniguchi, Role of OH radical in Fuel-NO_x formation during cocombustion of ammonia with hydrogen, methane, coal, and biomass. Energy Fuels (2020). https://doi.org/10.1021/acs.energyfuels.0c00356
5. W. Weng, L. Zhongshan, P. Marshall, P. Glarborg,Participation of alkali and sulfur in ammonia combustion chemistry: investigation for ammonia/solid fuel co-firing applications. Combust Flame (2022). https://doi.org/10.1016/j.combustflame.2022.112236
6. P. Chen, B. Jiang, H. Wang, M. Gu, Y. Fang, P. Wang, Experimental and theoretical calculations study on heterogeneous reduction of NO by char/NH_3 in the reduction zone of ammonia co-firing with pulverized coal: influence of mineral Fe. Fuel (2022). https://doi.org/10.1016/j.fuel.2021.122374
7. S. Ishihara, J. Zhang, T. Ito, Numerical calculation with detailed chemistry on ammonia co-firing in a coal-fired boiler: effect of ammonia co-firing ratio on NO emissions. Fuel (2020). https://doi.org/10.1016/j.fuel.2020.117742
8. J. Zhu, X. Liu, Y. Xu, J. Xu, H. Wang, K. Zhang, X. Cheng, D. Yu, Probing into volatile combustion flame and particulate formation behavior during the coal and ammonia co-firing process. Energy Fuels (2021). https://doi.org/10.1021/acs.energyfuels.2c01450

9. X. Wang, W. Fan, J. Chen, G. Feng, X. Zhang, Experimental study and kinetic analysis of the impact of ammonia co-firing ratio on products formation characteristics in ammonia/coal co-firing process. Fuel (2022). https://doi.org/10.1016/j.fuel.2022.125496
10. Y. Xu, H. Wang, X. Liu, J. Zhu, Z. Xu, M. Xu, Mitigating CO_2 emission in pulverized coal-fired power plant via co-firing ammonia: a simulation study of flue gas streams and exergy efficiency. Energy Convers. Manag. (2022). https://doi.org/10.1016/j.enconman.2022.115328
11. P. Chen, Y. Fang, P. Wang, M. Gu, K. Luo, J. Fan, The effect of ammonia co-firing on NO heterogeneous reduction in the high-temperature reduction zone of coal air-staging combustion: experimental and quantum chemistry study. Combust. Flame (2022). https://doi.org/10.1016/j.combustflame.2021.111857
12. S. Ishihara, J. Zhang, T. Ito, Numerical calculation with detailed chemistry of effect of ammonia co-firing on NO emissions in a coal-fired boiler. Fuel (2020). https://doi.org/10.1016/j.fuel.2019.116924
13. M.N. Rahman, M.A. Wahid, Renewable-based zero-carbon fuels for the use of power generation: a case study in Malaysia supported by updated developments worldwide. Energy Rep. (2021). https://doi.org/10.1016/j.egyr.2021.04.005
14. M.N. Rahman, M.H. Ujir, M.A. Wahid, M.F.M. Yasin, A single-step chemistry mechanism for biogas supersonic combustion velocity with nitrogen dilution. J. Therm. Anal. Calorim. (2022). https://doi.org/10.1007/s10973-022-11356-x
15. M.N. Rahman, N. Shahril, S. Yusup, Hydrogen-enriched natural gas swirling flame characteristics: a numerical analysis. CFD Lett. (2022).https://doi.org/10.37934/cfdl.14.7.100112
16. P. Chen, H. Wang, B. Jiang, Y. Wang, M. Gu, G. Chen, X. Huang, An experimental and theoretical study of NO heterogeneous reduction in the reduction zone of ammonia co-firing in a coal-fired boiler: influence of CO. Fuel Process Technol. (2022). https://doi.org/10.1016/j.fuproc.2022.107184
17. J.S. Cardoso, V. Silva, D. Eusebio, L.A.C. Tarelho, M.J. Hall, A.G. Dana, Numerical modelling of ammonia-coal co-firing in a pilot-scale fluidized bed reactor: influence of ammonia addition for emissions control. Energy Convers. Manag. (2022). https://doi.org/10.1016/j.enconman.2022.115226
18. M. Stocks, R. Fazeli, L. Hughes, F.J. Beck, Global emissions implications from co-combusting ammonia in coal fired power stations: an analysis of the Japan-Australia supply chain. J Clean Prod. (2022). https://doi.org/10.1016/j.jclepro.2021.130092
19. A. Valera-Medina, H. Xiao, M. Owen-Jones, W.I.F. David, P.J. Bowen, Ammonia for power. Prog. Energy Combust. Sci. (2018). https://doi.org/10.1016/j.pecs.2018.07.001
20. M. Tamura, T. Gotou, H. Ishii, D. Riechelmann, Experimental investigation of ammonia combustion in a bench scale 1.2 MW-thermalpulverised coal firing furnace. Appl. Energy (2020). https://doi.org/10.1016/j.apenergy.2020.115580
21. H. Lee, M. Lee, Recent advances in ammonia combustion technology in thermal power generation system for carbon emission reduction. Energies (2021). https://doi.org/10.3390/en14185604
22. M.N. Rahman, N.F. Othman, A numerical model for ash deposition based on actual operating conditions of a 700 MW coal-fired power plant: validation feedback loop via structural similarity indexes (SSIMs). CFD Lett. (2022). https://doi.org/10.37934/cfdl.14.1.99111
23. M.N. Rahman, Optimisation of solid fuel in-furnace blending for an opposed-firing utility boiler: a numerical analysis. CFD Lett. (2022).https://doi.org/10.37934/cfdl.14.9.89107
24. W. Sun, W. Zhong, T. Echekki, Large eddy simulation of non-premixed pulverized coal combustion in corner-fired furnace for various excess air ratios. Appl. Math. Model. (2019). https://doi.org/10.1016/j.apm.2019.05.017
25. K. Wan, J. Xia, Z. Wang, M. Pourkashanian, K. Cen, Large-eddy simulation of pilot-assisted pulverized-coal combustion in a weakly turbulent jet. Flow Turbul. Combust. (2017). https://doi.org/10.1007/s10494-017-9817-y

26. W. Sun, W. Zhong, J. Zhang, T. Echekki, Large eddy simulation on the effects of coal particles size on turbulent combustion characteristics and NO_x formation inside a corner-fired furnace. J. Energy Resour. Technol. (2021). https://doi.org/10.1115/1.4048864
27. V.B. Silva, C. João, *Computational Fluid Dynamics Applied to Waste to Energy Processes: A Hands-on Approach* (Butterworth-Heinemann, 2020)

Author Index

M. B. Othman et al. (eds.), *Proceedings of the 1st International Conference of New Energy*, Springer Proceedings in Energy, https://doi.org/10.1007/978-981-99-0859-2